U0149942

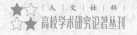

人｜文｜社｜科
高校学术研究论著丛刊

简要纺织工程技术专业英语

李 勇 吴 蓓 李 健 陈嘉琳 编著

中国书籍出版社
China Book Press

图书在版编目 (CIP) 数据

简要纺织工程技术专业英语 / 李勇等编著 . –– 北京：
中国书籍出版社 , 2022.7

ISBN 978–7–5068–9063–2

Ⅰ . ①简… Ⅱ . ①李… Ⅲ . ①纺织工业 – 英语 Ⅳ .
① TS1

中国版本图书馆 CIP 数据核字（2022）第 110581 号

简要纺织工程技术专业英语

李 勇 吴 蓓 李 健 陈嘉琳 编著

丛书策划	谭 鹏 武 斌
责任编辑	宋 然
责任印制	孙马飞 马 芝
封面设计	东方美迪
出版发行	中国书籍出版社
地　址	北京市丰台区三路居路 97 号 (邮编：100073)
电　话	（010）52257143（总编室） （010）52257140（发行部）
电子邮箱	eo@chinabp.com.cn
经　销	全国新华书店
印　厂	三河市德贤弘印务有限公司
开　本	710 毫米 × 1000 毫米　1/16
字　数	336 千字
印　张	18.75
版　次	2023 年 3 月第 1 版
印　次	2023 年 3 月第 1 次印刷
书　号	ISBN 978–7–5068–9063–2
定　价	88.00 元

目　录

Lesson 1   Cotton

Cotton fibers are the form of cellulose, nature's most abundant polymer. Nearly 90% of the cotton fibers are cellulose. All plants consist of cellulose, but to varying extents. Bast fibers, such as flax, jute, ramie and kenaf, from the stalks of the plants contain about threequarters of cellulose. The cellulose in cotton fibers is also of the highest molecular weight among all plant fibers and highest structural order, i.e., highly crystalline, oriented.

1.Chemical composition

Cotton fibers are composed of mostly α-cellulose (88.0%–96.5%). The non-cellulosics are located either on the outer layers (cuticle and primary cell wall) or inside the lumens of the fibers whereas the secondary cell wall is purely cellulose. The specific chemical compositions of cotton fibers vary by their varieties, growing environments (soil, water, temperature, pest, etc.) and maturity. The non-cellulosics include proteins (1.0%–1.9%), waxes (0.4%–1.2%), pectins (0.4%–1.2%), inorganics (0.7%–1.6%), and other substances (0.5%–8.0%). In less developed or immature fibers, the non-cellulosic contents are much higher.

2.Cellulose chemistry

Cotton cellulose is highly crystalline and oriented. α-cellulose is distinct in its long and rigid molecular structure. Cellulose is readily attacked by oxidizing agents, such as chlorous, peroxides. Most oxidizing agents are not selective in the way they react with the primary and secondary hydroxyl groups. Oxidation of cellulose can lead to two products, reducing and acidic oxycellulose. Heating cellulose up to 120 ℃ drives off moisture without affecting its strength. Heating to a higher 150℃ has been shown to reduce molecular weight, and tensile the strength. Drying of fibers involves the removal of fluids from the lumens and intermolecular water in the cellulose. The fluid loss from the lumens causes the cylindrical fibers to collapse to form twists or convolutions. The loss of intermolecular water allows the cellulose chains to come closer together and form intermolecular hydrogen bonds. Prior to ball dehiscence and fiber desiccation, matured cotton fibers have been shown to exhibit high porosity in their structure. The accessibility of water in fiber structure in the hydrated state is higher than after desiccation.

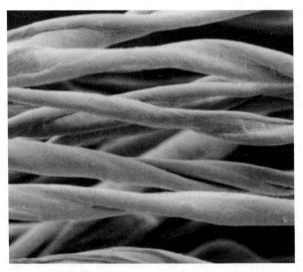

Figure 1.1　Scanning electron micrographs of mature fibers

The matured fibers (see Fig. 1.1) dry into flat twisted ribbon forms. The twist or convolution directions reverse frequently along the fibers. The number of twists in cotton fibers varies between 3.9 and 6.5 per mm and the spiral reversal changes one to three times per mm in length. Both the structure and compositions of the cellulose and non-cellulosics depend on the variety and the growing conditions. The strength of Cotton fibers has much to do with both the primary wall formation and the secondary wall thickening as well as during desiccation. The overall crystallinity and the apparent crystallite sizes increase with the fiber development for cotton fibers.

生词与词组

1.Cotton fiber 棉纤维

2.cellulose ['seljuləʊs]*n.* 纤维素

3.polymer ['pɒlɪmə(r)]*n.* 聚合物

4.bast fiber 韧皮纤维

5.flax [flæks]*n.* 亚麻

6.jute [dʒu:t]*n.* 黄麻

7.ramie ['ræmɪ]*n.* 苎麻

8.kenaf ['kɛnæf]*n.* 洋麻

9.molecular [mə'lekjələ(r)]*adj.* 分子的；分子内的

10.crystalline ['krɪstəlaɪn]*adj.* 结晶的

11.oriented ['ɔ:rientɪd]*adj.* 取向的

12.cuticle ['kju:tɪkl]*n.* 表皮；角质层

13.lumen ['lu:mɪn]*n.* 管腔；中腔

14.pest [pest]*n.* 害虫

15.maturity [mə'tʃʊərəti]*n.* 成熟度；成熟作用

16.cellulosic [seljʊ'ləʊsɪk]*adj.* 纤维素的；纤维素质的

17.protein ['prəʊti:n]*n.* 蛋白；蛋白质；朊

18.wax [wæks]*n.* 蜡

19.pectin ['pɛktɪn]*n.* 果胶

20.inorganic [ˌɪnɔ:'gænɪk]*adj.* 无机的

21.substance ['sʌbstəns]*n.* 物质

22.oxidizing agent 氧化剂

23.chlorous ['klɔ:rəs]*adj.* 次氯酸的；次氯的

24.peroxide [pə'rɒksaɪd]*n.* 过氧化物

25.hydroxyl group 羟基；羟基团

26.oxidation of cellulose 氧化纤维素

27.acidic oxycellulose 酸性氧化纤维素

28.moisture ['mɔɪstʃə(r)]*n*. 水分；湿气；水气

29.tensile strength 拉伸强度；抗拉强度

30.fluid loss 失水；失水量

31.collapse [kə'læps]*n*. 塌陷；崩塌；坍塌

32.twist [twɪst]*n*. 扭转；扭角；捻

33.convolution [ˌkɒnvə'luːʃ(ə)n]*n*. 转曲；回旋

34.intermolecular water 分子间水

35.intermolecular hydrogen bond 分子间氢键

36.ball dehiscence 棉铃裂开

37.fiber desiccation 纤维干扁；纤维干化

38.porosity [pɔː'rɒsəti]*n*. 孔隙率

39.hydrated state 水合状态；水化状态

40.flat twisted ribbon 扁扭带

41.spiral ['spaɪrəl]*adj*. 螺旋形的；螺旋式的

译文

第 1 课　棉

棉纤维由自然界最丰富的聚合物——纤维素形成。棉纤维 90% 的成分为纤维素。所有植物都由纤维素组成，但含量不同。来自植物茎的韧皮纤维（如亚麻、黄麻、苎麻和洋麻），纤维素含量约为四分之三。棉纤维的纤维素分子的分子量高，其结构排列高度有序（如高结晶度、高取向度）。

1. 化学成分

棉纤维主要由 α- 纤维素（88.0%-96.5%）组成。非纤维素物质位于外层（表层和初生胞壁）或纤维腔内，而次生胞壁是纯纤维素。棉花纤维的化学成分因品种、生长环境（土壤、水、温度、虫害等）和成熟度而异。非纤维素物质包括蛋白质（1.0%-1.9%）、蜡质（0.4%-1.2%）、果胶（0.4%-1.2%）、无机物（0.7%-1.6%）和其他物质（0.5%-8.0%）。生长不完全或未成熟纤维的非纤维素含量更高。

2. 纤维素化学性

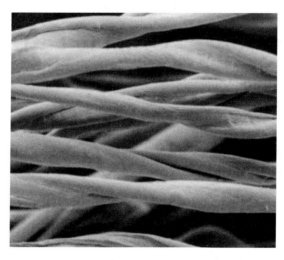

图 1.1   成熟棉纤维电镜图

棉纤维素具有高度结晶性和取向性。α- 纤维素的独特之处在于其长而坚硬的分子结构。纤维素易与氧化剂反应,如次氯酸、过氧化物。大多数氧化剂对伯羟基和仲羟基的化学反应没有选择性。纤维素的氧化产物为还原性和酸性氧化纤维素。纤维素加热至 120℃可去除水分,并且温度对强度无影响。加热至 150℃纤维素的分子量和拉伸强度均降低。纤维干燥包括空腔内水分的去除及纤维素间分子水的去除。管腔内流体的散失引起了圆柱形纤维的扭转。分子间水的散失使纤维素分子链靠近并形成分子间氢键。苞(棉铃)开裂、纤维干化前,成熟的棉纤维已展现了高孔隙率结构。水合状态下的纤维结构比干态纤维更亲水。

成熟纤维(如图 1.1)变干转曲成扁平带状。转曲或回旋方向沿纤维轴频繁换向。棉纤维每毫米转曲 3.9-6.5 次,每毫米螺旋反转 1-3 次。纤维素和非纤维素的结构和组成都取决于棉品种和生长条件。棉纤维强度与初生层和次生层以及其干燥过程都有很大关系。棉纤维的整体结晶度和表观晶粒尺寸随纤维的生长而增加。

Lesson 2   Organic cotton

Cotton grown without the use of any synthetically compounded chemicals (i.e., pesticides, plant growth regulators, defoliants, etc.) and fertilizers is considered "organic" cotton. However, chemicals considered "natural" can be used in the production of organic cotton as well as natural fertilizers. "Organic" is a labeling term. For cotton to be sold as "organic cotton", it must be certified by an independent organization that verifies that it meets or exceeds defined organic agricultural production standards. To produce "organic cotton textiles", certified organic cotton should be manufactured according to organic fiber processing standards/guidelines. A three-year transitional period from conventional to organic cotton production is required for certification. Cotton produced during this three-year period is described variously as "transitional", "pending certification", or "organic B". There continues to be worldwide interest in the organic cotton as a potentially environmentally friendly way to produce cotton and for economic reasons. Production of organic cotton has increased recently to about 0.1% of world cotton production, mainly due to increased production in Turkey as well as India and some African countries. Organic production is not necessarily any more environmentally friendly or sustainable than current conventional cotton production. From a consumer standpoint of residue, there is no difference between conventionally grown cotton and organically grown cotton. Growing organic cotton is more demanding and more expensive than growing cotton conventionally. Organic production can be a real challenge if pest pressures are high. Conventional and organic cotton production can co-exist.

生词与词组

1.synthetically compounded chemical 合成化学品
2.pesticide [ˈpestɪsaɪd]*n.* 杀虫剂；农药
3.plant growth regulator 植物生长调节剂
4.defoliant [diːˈfəʊliənt]*n.* 落叶剂
5.fertilizer [ˈfɜː(r)təlaɪzə(r)]*n.* 化肥；精媒介物
6.pending certification 待认证
7.organic cotton 有机棉
8.co-exist [kəʊˌɪgzɪst]*v.* 共存；并存

译文

第 2 课　有机棉

　　棉花种植不使用任何合成化学品(即杀虫剂、植物生长调节剂、落叶剂等)和化肥,被认为是"有机"棉花。然而,天然肥料等"天然"的化学品可用于生产有机棉。"有机"是一个标签术语。作为"有机棉"出售的棉花,应由第三方机构认证,以证明其符合或高于有机农业生产规定标准。认证的有机棉应按照有机纤维加工标准 / 指南,生产"有机棉纺织品"。常规棉花认证为"有机棉"需要三年时间。在三年期内,待认定的棉花类型被描述为"过渡""待认证"或"有机 B"。出于经济原因,有机棉作为一种潜在的环境友好型棉花生产方式受到全世界的关注。

　　有机棉的产量已增加到世界棉花产量的 0.1%,这归因于土耳其、印度及一些非洲国家的有机棉增产。有机生产不一定比目前的常规棉花生产更环保或可持续。从消费者对残留物的观点来看,常规种植的棉花和有机种植的棉花之间并无区别。种植有机棉比种植常规棉花要求更高,种植成本也更高。如果虫害严重,对有机生产是一个真正的挑战。常规棉和有机棉的生产应并存。

Lesson 3 The harvesting and ginning of cotton

Cotton possesses its highest fiber quality and best potential for spinning when the bolls are mature and freshly opened. Quality of the fiber in the bale depends on many factors, including variety, weather conditions, cultural practices, harvesting and storage practices, moisture and trash content, and ginning processes. Genetics plays an important role in fiber quality, both in the initial quality of the fiber as well as how well the fiber withstands gin processes.

Figure 3.1   Mechanical harvester for cotton

Harvesting practices from hand-picked to machine-stripped(see Fig. 3.1) dramatically impact the amount of trash entangled with the cotton and thus the amount of cleaning machinery required at the gin. Fiber quality factors such as length, uniformity, micronaire, strength, short fiber content, neps, and seedcoat fragments may differ dramatically for varieties grown under nearly identical conditions. The colour is substantially affected by weather and length of exposure to weather conditions after the bolls open. Abnormal colour (light-spot, spotted, tinged, yellow-stained, etc.) indicates a deterioration in quality. Continued exposure to weather and the action of micro-organisms can cause white cotton to lose its brightness and become darker in colour. The weakened fibers cannot withstand the standard lint-seed separation or lint cleaning processes without additional damage and fiber loss. In fact, varieties and excessive weathering have a far greater impact on fiber quality than the most rigorous of gin processes. Cotton gins are responsible for converting a raw agricultural product, seed cotton, into commodities such as bales of lint, cottonseed, motes, compost, etc. Gins are a focal point of the cotton community and their location, resources, and contributions to the economy are critical to the cotton industry.

Enormous differences exist across the worldwide spectrum of cotton production, harvesting and ginning. Harvest methods range from totally hand harvested in some countries to totally machine-harvested in others. In fact, only the United States and Australia are fully mechanized. Cotton storage after harvesting ranges from small piles of cotton on the ground in some countries to mechanically made modules containing over 12 tons of cotton in others. Production of high quality cotton begins with the selection of varieties and continues through the use of good production practices that include harvesting, storage, and ginning. Gins can preserve fiber quality and dramatically improve market grade and value. The principal function of the cotton gin is to separate lint from seed cotton. The gin then must also be equipped to remove a large percentage of the foreign matter from the cotton that would significantly reduce the value of the ginned lint, especially if the cotton is machine harvested. A ginner must have two objectives: first, to produce lint of satisfactory quality for the grower's classing and market system, and secondly, to gin the cotton with minimum reduction in fiber spinning quality so that the cotton will meet the demands of its ultimate users, the spinner and the consumer. Thus, quality preservation during ginning requires the proper selection and operation of every machine that is included in a ginning system. The ginner must also consider the weight loss that occurs in the various cleaning machines.

生词与词组

1.spinning [ˈspɪnɪŋ]*n.* 纺纱 *adj.* 纺纱的

2.boll [bəʊl]*n.* 棉铃；棉桃

3.bale [beɪl]*n.* 大包；大捆

4.harvesting [ˈhɑː(r)vɪstɪŋ]*n.* 收获；收割

5.trash content 杂质含量；含杂率

6.ginning process 轧花过程

7.genetics [dʒəˈnetɪks]*n.* 遗传学；遗传特征

8.gin [dʒɪn]*n.* 轧棉机；弹棉机

9.hand-picked ['hænd'pɪkt]*adj.* 手摘的；手工挑选的

10.uniformity [ˌjuːnɪˈfɔː(r)məti]*n.* 整齐度；均匀性；均匀度；一致性

11.micronaire [ˈmaɪkrɔneə(r)]*n.* 马克隆值

12.nep [nep]*n.* 绵结；白星；毛粒

13.seedcoat fragment 破籽

14.abnormal colour 颜色异常

15.spotted [ˈspɒtɪd]*adj.* 有斑点的；玷污的；有污点的

16.tinged [tɪndʒd]*adj.* 淡色的；略微的

17.weathering [ˈweðə(r)ɪŋ]*n.* 风化；大气；老化；大气暴露

18.seed cotton 籽棉

19.cottonseed [ˈkɒtənsiːd]*n.* 棉籽

20.mote [məʊt]*n.* 尘屑；微尘

21.compost [ˈkɒmpɒst]*n.* 混合肥料；堆肥

22.spectrum [ˈspektrəm]*n.* 光谱；系列

23.hand harvested 手工采摘的

24.machine-harvested [məˈʃiːn ˈhɑːvɪst]*adj.* 机器采摘的

25.foreign matter 杂质；异物；异性纤维

26.spinner [ˈspɪnə(r)]*n.* 纺纱工

译文

第3课　棉花收获与加工

当棉铃成熟开裂时,棉花具有最高的纤维质量和极佳的可纺性。棉包中纤维的质量取决于多种因素,包括品种、气候条件、栽培技术、收获和储存方法、含水量和杂质含量以及轧花工艺。在纤维初始质量和纤维轧花承受能力方面,纤维质量主要由基因决定,也受纤维初始质量和轧花耐受性影响。

图 3.1　采棉机

从手工采摘到机器采摘(见图 3.1)的收获方式极大地影响了缠结在棉花上的杂质含量,从而决定了轧花环节所需的清洁机械数量。在几乎同等生长条件下的棉花品种,其纤维质量因素(如长度、整齐度、马克隆值、强力、短绒含量、棉结、破籽)存在很大差异。棉铃开裂后,棉花颜色在很大程度上受天气和暴露外界环境时间长短的影响。异常颜色(亮点、斑点、淡色、黄染等)表示棉花质量下降。持续暴露在空气和微生物的作用下会导致白棉失去色泽而颜色变暗淡。变脆弱的纤维无法承受标准的绒籽分离或棉绒清洁过程,必然会造成额外的损坏和纤维损失。事实上,与最严格的轧花过程相比,品种和过度风化对纤维质量的影响要大得多。轧棉机对未加工的农产品(籽棉)进行轧制,其转化为大包的皮棉、棉籽、微粒、堆肥等商品。轧花机是棉花产业区的焦点,其所处的位置、资源及其经济贡献对棉花产业至关重要。

　　全世界的棉花生产、收获和轧花存在巨大差异。从一些国家的全手工收获到部分国家的全机器收获,实际上只有美国和澳大利亚的收获方法是完全机械化的。收获后,棉花储存包括一些国家的地面小堆堆棉至部分国家的机械打模(模包重量达 12 吨)。优质棉花的生产始于品种的选择,再经收获、储存和轧花等环节实现优质生产。轧花可保持纤维质量,并可显著提升市场棉花的等级及价值。轧棉机是将皮棉从籽棉中分离出来。轧棉机还必须配备能从棉花中去除大量异性纤维的装置,其异性纤维会显著降低轧花皮棉的价值,机械采摘的棉花更是如此。轧花厂有两个目标:首先,生产出符合种植者分级和市场系统质量的皮棉;其次,以最小的纤维纺制质量下降为原则对棉纤维进行轧花,从而使棉花满足最终用户——纺纱工和消费者的需求。因此,在轧花过程中合理选择轧花系统,合理操作每一台机器对纤维质量至关重要。保持质量需要正确选择和操作轧花系统中包含的每台机器。轧花厂也需考虑棉纤维在清理机械中的损失。

Lesson 4　Jute

Among the natural fibres, jute ranks next to cotton in terms of production. Jute (see Fig. 4.1) is a cellulosic fibre under the category of bast fibres and its cultivation is almost as old as human civilization. Jute, an annual herbaceous plant, is mainly cultivated in South and South East Asia. Jute was first used as an industrial raw material for making packaging materials, replacing flax and hemp grown in Europe.

Figure 4.1　Jute fibres

The harvesting time of jute is calculated by taking into account of the crop age, height and flowering stage, which varies according to the grown and sowing time of the varieties, but usually 110–120 days are required for jute to mature for harvesting.

After harvesting and defoliation of plants in the fields for 3–4 days, the jute stems are retted in water and the fibre is extracted. The traditional method is to ret the jute stems for about 15–18 days and extract the fibre manually after retting. After retting, plants are taken out of water and the fibre is traditionally extracted by hand.

The washed fibre is spread over a bamboo bar for thorough sun drying for 4–7 days before storage. The main parameters of fibre quality include colour, lustre, strength, texture, length, etc.

Jute fibre, unlike cotton, is a multicellular fibre. In the jute plant the fibre is formed as a cylindrical sheath made up of single fibres (ultimate cell) joined together in such a way as to form a three-dimensional network from top to bottom of the stem.

Jute fibre is basically a compound of lignocellulose. It is a complex of organic molecules, which on combustion leaves a little ash consisting of calcium, magnesium, aluminium, iron, etc., that are present either in the free state or bonded with functional groups of cellulosic chain. The number of ultimate cells in one such bundle constituting a single fibre ranges from a minimum of 8–9 to a maximum of 20–25. This wide variation in the number of cells in a bundle is believed to be a major cause of variation in the physical and mechanical properties of the fibre and its quality.

The spinnable units in jute fibre strands are, like those in most other bast fibre crops, filaments composed of a string of cells bonded together by pectin and hemicelluloses.

Commercial jute ranges from pale cream to golden yellow and from light brown to dirty grey in colour. It possesses a natural silky shine. Jute is a relatively coarse, stiff, inelastic and somewhat rigid fibre that has slightly higher moisture regains (12%–13%) than cotton (7%–8%). Good frictional property, tenacity, very high modulus and low breaking elongation make jute an ideal packaging material.

The use of jute is limited to coarse fabrics, because the length/diameter ratio of jute filaments is only 100–120, which is much below the minimum of 1,000 required for fine spinning quality. Jute fibre is hygroscopic and wetted filaments may swell up to 23% in diameter. Other than being of agro-origin and biodegradable, the major advantageous features of jute are its high strength and initial modulus, moderate moisture regain, good dyeability using different dyes, good heat and sound insulation properties and low cost. However, the major disadvantages of jute are its coarseness, stiffness, low wet strength, moderate wash shrinkage, harsh feel, hairiness and high fibre shedding, photo-yellowing, and poor crease recovery.

Jute has been the most widely used packaging fibre for more than 100 years because of its strength and durability, low production costs, ease of manufacturing. Other traditional products include hessian cloth, carpet backing, yarn, twine, cordage, nonwoven felts.

生词与词组

1.bast fibre 韧皮纤维

2.annual herbaceous plant 一年生草本植物

3.packaging material 包装材料

4.hemp [hemp]*n.* 大麻；大麻纤维

5.flowering [ˈflaʊərɪŋ]*adj.* 开花的

6.sowing [ˈsəʊɪŋ]*n.* 播种

7.defoliation [ˌdiːˌfəʊlɪˈeɪʃ(ə)n]*n.* 落叶

8.stem [stem]*n.* (花草的) 茎；柄

9.ret [ret]*v.* 沤 (麻 , 肥料等)；浸；受潮湿腐烂

10.extract [ɪkˈstrækt]*v.* 提取；萃取

11.multicellular fibre 多细胞纤维

12.cylindrical sheath 圆柱鞘；圆柱护皮

13.three-dimensional network 三维网

14.lignocellulose [ˌlɪgnəʊˈseljʊləʊs]*n.* 木质纤维素

15.combustion [kəmˈbʌstʃ(ə)n]*n.* 燃烧；发火；点火

16.calcium [ˈkælsiəm]*n.* 钙

17.magnesium [mægˈniːziəm]*n.* 镁

18.spinnable [ˈspɪnəbl]*adj.* 可织的；适织的

19.pectin [ˈpektɪn]*n.* 果胶；胶质

20.hemicellulose [ˌhemɪˈseljʊləʊs]*n.* 半纤维素

21.pale cream 淡奶油色

22.natural silky shine 天然丝光

23.coarse [kɔ:(r)s]*adj.* 粗的

24.inelastic [ˌɪnɪˈlæstɪk]*adj.* 无弹性的；无弹力的；非弹性的

25.frictional [ˈfrɪkʃənəl]*adj.* 摩擦的

26.tenacity [təˈnæsəti]*n.* 韧性；坚韧度

27.breaking elongation 断裂伸长；断裂伸长率

28.swell up 润胀；膨胀

29.biodegradable [ˌbaɪəʊdɪˈgreɪdəb(ə)l]*adj.* 可生物降解的

30.dyeability [daɪəˈbɪləti]*n.* 染色性能；可染性；上染率；染色度

31.insulation [ˌɪnsjʊˈleɪʃ(ə)n]*n.* 保温；隔热

32.stiffness [ˈstɪfnəs]*n.* 刚度；硬挺度；抗弯刚度

33.wash shrinkage 洗涤收缩，洗水皱缩；洗涤缩水

34.hairiness [ˈheərɪnəs]*n.* 毛羽；有毛

35.fibre shedding 纤维脱落

36.photo-yellowing 发黄；光变黄

37.poor crease recovery 折痕恢复差

38.hessian cloth 粗麻布；麻制粗布

39.carpet backing 地毯背衬

40.yarn [jɑ:(r)n]*n.* 纱；毛线

41.twine [twaɪn]*n.* 麻线

42.cordage [ˈkɔ:dɪdʒ]*n.* 绳索；缆索

译文

第4课　黄麻

在天然纤维中,黄麻的产量仅次于棉花。黄麻属于韧皮纤维类的纤维素纤维,其栽培历史几乎与人类文明一样古老。黄麻(见图4.1)是一年生草本植物,主要种植在南亚和东南亚。黄麻取代了欧洲种植的亚麻和大麻,其被用作制造包装材料的工业原料。

图 4.1　黄麻纤维

黄麻的收获时间是根据作物的种植时间、高度和花期计算的,因种植品种和播种时间而异,但通常黄麻需要 110–120 天才能成熟收获。在收获和脱叶 3–4 天后,黄麻茎浸入水沤麻并提取纤维。传统的方法是将黄麻茎沤约 15–18 天,再手工提取纤维。

洗涤后的纤维铺在竹条上,在储存前用 4–7 天彻底晒干。纤维质量的主要参数包括颜色、光泽、强度、纹理、长度等。

与棉花不同,黄麻纤维是一种多细胞纤维。黄麻植物的纤维是由单纤维组成的圆柱鞘,从茎的顶部到底部以三维网络的方式连接在一起。

黄麻纤维基本上是木质纤维素的化合物。它是一种有机分子的化合物,其燃烧会产生少量灰烬,灰烬由钙、镁、铝、铁等组成,它们以游离状态存在或与纤维素链的官能团结合。

构成纤维束的单根纤维数量范围从最少 8–9 根到最多 20–25 根。纤维束中细胞数量的变化被认为是纤维物理和机械性能及其质量变化的主要原因。黄麻纤维束中的可纺单元与大多数其他韧皮纤维作物中的可纺单元一样,是由果胶和半纤维素黏合在一起的一串细胞组成的细丝。

商业黄麻的颜色范围从淡奶油色到金黄色,从浅棕色到暗灰色。它具有天然的丝光泽。黄麻是一种相对粗糙、坚硬、无弹性且有些刚性的纤维,其回潮率(12%–13%)略高于棉花(7%–8%)。良好的摩擦性能、韧性、极高的模量和低断裂伸长率使黄麻成为理想的包装材料。

　　黄麻的使用仅限于粗纺织物,因为黄麻长丝的长径比仅为100-120,远低于精纺质量所要求的最低1000的数值。黄麻纤维具有良好吸湿性,湿态丝直径可膨胀23%。除了天然和可生物降解之外,黄麻的优点是其强度和初始模量高、回潮率适中、染色性良好、隔热和隔音效果良好、成本低。然而,黄麻的缺点是粗糙、僵硬、湿强度低、洗涤收缩性适中、手感粗糙、毛羽和高纤维脱落率、发黄和折皱恢复性差。

　　黄麻因其强度高和耐用、成本低、易于制造,已用于包装材料100多年了。其他传统产品包括粗麻布、地毯背衬、纱线、麻线、绳索、无纺毡。

Lesson 5   Ramie

Ramie is one of the oldest fibre crops, having been used for at least 6,000 years. Ramie fibre is one of the strongest and longest natural fine textile fibres in the world. It is a bast fibre derived from the bast layer of the stem, that is, phloem of the vegetative stalks of the plants. China, mainly the central and southern part, leads the world in the production of ramie. Other major producers of ramie fibre are Japan, Brazil, Indonesia and India. Ramie, being a perennial plant, is normally harvested two to three times a year, but under favourable growing conditions it can be harvested up to six times per year. The income generally starts from the second year and continues thereafter. Ramie has been proved quite remunerative when grown under favourable edaphic and climatic conditions.

Ramie fibre generally is graded according to length, colour and cleanliness. Top grades usually are washed and sometimes brushed. There is no standard set of grades for ramie fibre, but several countries have set up their own grades.

Ramie is a multicellular bast fibre, by and large cellulosic in nature, having very little lignin and hemicellulose. The intercellular binding constituents present in significant amounts are natural gums and pectinous matters. The cells of ramie fibre may be as long as 40cm–45cm, cylindrical in nature and characterized by thick walls and narrow, curved lumens. The raw ramie fibre strand has an average length of 0.61m–1m. The longer fibres are sometimes more than 1.5m–2m in length. These are not single fibres, rather a bundle of shorter single fibres, as in other bast fibres, held together by gummy and pectinous matters. The elementary cells/single fibres of ramie are longer and thicker than all other bast fibres. Coarse ramie fibres are generally used for making twines and threads, for which its strength and lack of stretch make it most suitable. Because of its high wet strength, quick dry ability and considerable resistance to bacterial action, it is very useful for making fishing nets. Ramie is used in many diverse applications like suiting, shirting, sheeting, dress materials, table cloths, napkins, towels, handkerchiefs, fine furniture upholstery, draperies, mosquito netting, gas mantles, industrial sewing thread, packing materials, fishing nets, fire hose, belting, canvas, marine shaft packing, knitting yarns, hat braids, filter cloths, etc.

生词与词组

1.bast layer 韧皮层

2.phloem [ˈfləʊem]n. 韧皮部

3.perennial [pəˈreniəl]adj. 多年生的 n. 多年生植物

4.remunerative [rɪˈmju:nərətɪv]adj. 有报酬的；有利的

5.edaphic climatic condition 风土条件；气候条件

6.brushed [brʌʃt]adj. 拉绒的；起绒的

7.lignin [ˈlɪgnɪn]n. 木质素

8.intercellular binding constituent 细胞间结合成分

9.natural gum and pectinous matter 天然胶类和胶质类物质

10.elementary cell 初生细胞

11.thread [θred]n. 细线；细丝

12.suiting [ˈsu:tɪŋ]n. 套装

13.dress material 衣料

14.table cloth 桌布

15.napkin [ˈnæpkɪn]n. 餐巾（纸）

16.towel [ˈtaʊəl]n. 纸巾

17.handkerchief [ˈhæŋkə(r)ˌtʃɪf]n. 手帕

18.fine furniture upholstery 精美的家具装饰

19.drapery [ˈdreɪpəri]n. 服装；布匹；绸缎；呢绒

20.mosquito netting 蚊帐

21.gas mantle 气罩

22.industrial sewing thread 工业缝纫线

23.packing material 包装材料

24.fishing net 渔网

25.fire hose 灭火水龙带；救火蛇管

26.belting [ˈbeltɪŋ]*n.* 带；带料

27.canvas [ˈkænvəs]*n.* 帆布；帐篷

28.marine shaft packing 船用轴包装

29.knitting yarn 针织纱

30.hat braid 帽辫

31.filter cloth 滤布

译文

第5课　苎麻

　　苎麻是最古老的纤维作物之一,已经使用了至少6000年。苎麻纤维是世界上强度最大、长度最长的天然纺织纤维之一。它是一种源于茎的韧皮层韧皮纤维,即植物茎的韧皮部。中国的中部和南部出产苎麻,其产量居世界前列。其他生产苎麻纤维的国家包括日本、巴西、印度尼西亚和印度。苎麻是多年生植物,通常每年可收获2至3次,但在有利的生长条件下,每年最多可收获6次。苎麻的收获从第二年开始,并在此后继续。事实证明,在优越的土壤气候条件下种植苎麻是非常有利的。

　　苎麻纤维一般按照长度、颜色和整洁度进行分级。高级品苎麻通常需洗涤,有时也需刷洗。苎麻纤维尚无标准的分级,但一些国家已设定了自己的苎麻纤维分级。

　　苎麻是一种多细胞韧皮纤维,其本质上为纤维素纤维,几乎没有木质素和半纤维素。苎麻纤维细胞间存在大量天然胶和果胶物质。苎麻纤维的细胞可长达40-45厘米,细胞外形呈圆柱形的,细胞壁厚,管腔狭窄弯曲。苎麻原纤的平均长度为0.61-1米,长纤维甚至可达1.5-2米。这些不是单纤维,而是由短纤维组成的纤维束。和其他韧皮纤维一样,苎麻短纤维由胶质和果胶物质粘结成纤维束。苎麻的基本细胞(单纤维)比其他韧皮纤维更长更粗。粗苎麻纤维通常用于制造高强低伸的麻绳和线。因苎麻纤维湿强高、排湿快和抗菌,适用于制作渔网。

苎麻也可用于制作西装、衬衫、床单、服装材料、桌布、餐巾、毛巾、手帕、精美的家具内饰、窗帘、蚊帐、气罩、工业缝纫线、包装材料、渔网、消防水带、皮带、帆布、船用轴包装、针织纱、帽辫、滤布等。

Lesson 6　Flax and hemp

Various parts of the flax plant—such as seeds, leaves, bast and the woody core—have potential as a source of valuable textile fibres. Flax fibres, applied traditionally for textile utilization such as woven, knitting and technical textiles have been used for many centuries. Application of coarse flax fibres are: pulp and nonwovens.

Fibre production from hemp has been conducted over many centuries, for end uses from textiles, ropes and sails, to matrices for industrial products in the modern age. It has been cultivated and utilised in a diversity of countries in many parts of the world, in both the northern and southern hemisphere. Hemp is a plant from which bast fibres can be extracted from the stems of the plants.

The extraction of fibres from hemp stems is commonly achieved by the mechanical processing of the straw, although historically manual decortication has been widely practiced. The purpose of this processing stage is to completely extract the fibres by entirely separating them from the woody core of the hemp stem, which is often referred to as "hurd".

The quality and quantity of hemp fibres yielded from a crop is very strongly determined by the degree of retting that the fibre has undergone. The retting process involves the removal and breaking down of the "gummy" substances, particularly pectins. The retting process can be undertaken chemically or biologically. Prior to processing, after retting, it is necessary that the straw be dried to a suitable moisture content to stop the retting process and to enable the straw to be stored without deterioration. Conventionally, hemp fibres were extracted in a long fibre (50cm–60cm) form, with a significant quantity of by-product short fibre. However, as new markets have developed for hemp fibres, decortication methods of fibre extraction that aim to produce a single quality of hemp fibre have been developed.

生词与词组

1.woven ['wəʊvən]*n.* 机织；机织物；梭织 *adj.* 织物的

2.pulp [pʌlp]*n.* 纸浆；浆粕

3.nonwoven [ˌnɒnˈwəʊvən]*adj.* 非织造的；无织的

4.rope [rəʊp]*n.* 绳；粗绳；线缆；绳索

5.sail [seɪl]*n.* 帆；航行

6.matrix ['meɪtrɪks]*n.* 基质；矩阵

7.hemisphere ['hemɪˌsfɪə(r)]*n.*(地球的) 半球

8.extraction [ɪkˈstrækʃ(ə)n]*n.* 提取；抽出；拔出

9.manual decortication 手动剥皮；人工剥皮

10.substance ['sʌbstəns]*n.* 物质；材料；实体；本体

11.deterioration [dɪˌtɪərɪəˈreɪʃ(ə)n]*n.* 恶化；变质；退化

译文

第 6 课　亚麻和大麻

亚麻植物的各个部分(如种子、叶子、韧皮和木质部)都有可能成为有价值的纺织纤维原料。作为传统的机织物、针织物、功能性纺织品,亚麻纤维已被使用了几个世纪。亚麻粗纤维可用于纸浆和无纺布。

大麻纤维生产已经进行了几个世纪,其终端用途从纺织品、绳索、帆,到现代工业产品的基质材料。它在北半球和南半球的许多地区种植和使用。大麻是一种可从植物茎中提取韧皮纤维的植物。

以往,从大麻茎中提取纤维采用手工剥皮,现通常由机械加工完成。该加工是将纤维与大麻茎的木质部完全分离,以提取纤维,这通常被称为"剥麻"。

从大麻作物获得的纤维质量和数量很大程度上取决于纤维所经历的沤麻程度。沤麻过程包括去除和分解"黏性"物质,特别是果胶。浸渍过程可采用化学或生物方式进行。在沤制之后,秸秆需要干燥至合适的湿度,以便于储存而不变质。一般提取出来的大麻纤维为 50-60 厘米的长纤维,并带有大量短纤维。然而,随着新兴大麻纤维市场的发展,单一质量大麻纤维的纤维提取方法已开发。

Lesson 7   Wool

Sheep husbandry is an important pastoral activity across most of Europe, the Americas and Asia. Its purpose nowadays is primarily meat production, although the wool harvest has for centuries been an important basis of local textile industries.

1.Wool harvesting

Most wool is harvested by shearing live sheep using powered hand clippers. Blade shears are still preferred for flocks run where harsh weather can occur, because about 10mm of fleece can be left on the sheep for protection. Woolly sheepskins from slaughtered sheep and lambs are chemically treated to weaken the fibre roots so that the wool can be pulled off the pelt. The main wool classifications are full fleece (a year's growth, see Fig. 7.1), second shear and early shorn (part-year's growth), and crutchings (shorn from the hindquarters before lambing).

Figure 7.1   Full fleece of wool

2.Clip preparation

During shearing, each fleece is examined with the objective of removing faults relating to colour, length and contamination, and collecting each separately. The main categories are belly wool, short discoloured crutch wool, stains and so on. Vegetable matter and cotted (i.e. felted) parts of the fleece are also separated.

Fine wool lines will discriminate on the variables of fibre diameter, length, strength, colour and vegetable contamination. The main sorting principles for carpet wools are discolouration and fleece tenderness.

3.Wool scouring, carbonising

Raw or "greasy" wool is contaminated with impurities, the type depending on the breed of sheep, the area in which the sheep are raised, and husbandry methods.

The role of wool scouring is to:

• Clean the contaminants from the wool;

• Ensure that the wool is in a physical and chemical condition to suit the intended processing route (e.g. for topmaking to minimise entanglement and retain the staple structure).

The term "scouring" is used here in the generic sense of a process that removes contaminants from raw wool. Thus, it includes all processes which aim to clean wool including those which use solvents other than water and those which use solids as a carrier for removing the contaminants. Scouring clearly is a critically important step in wool processing.

Figure 7.2   Angora goat

4.Fibre diameter

No measurement could better exemplify the divergence in priorities between fine and coarse wools than measurement of fibre diameter. For fine Merino wools, the predominant average within a lot lies between 18 and 21 microns and there is a useful price premium at the finer end of that range. The mean diameter, of course, relates closely to both the spinning limit and the luxury handle, where every micron finer creates an advantage.

5.Fibre morphology

Wool is the generally accepted generic description of the hair of various breeds of domesticated sheep, although it is also commonly used as the generic name of all animal hair, particularly including the so-called fine animal hair, i.e. the hair of the cashmere and Angora goat (Angora goat, see Fig. 7.2), of camel, vicuna and alpaca, of the Angora rabbit, and of the yak. The morphology and composition of human hair also closely resembles that of wool. While wool contains α-keratins, silk and feathers are composed of β-keratins. From a macromolecular point of view, wool is a composite fibre, i.e. a fibril-reinforced matrix material with both the fibrils and the matrix consisting of polypeptides, interconnected physically and chemically. From a morphological point of view, the wool fibre is a nanocomposite of high complexity with a clear hierarchy indicating an enormous degree of self-organisation.

6.Wool chemistry

Over the full range of wool types, fibre diameters vary between about 10mm and 80mm. Diffusion of reactant chemicals from an immersion solvent, which most commonly is water, may be slow. In addition to fibre diameter variations, diffusion is also moderated by the hydrophobic epicuticle on the external face of wool scale cells. Within each cell there are more variations in protein organisation, resulting in micro-heterogeneous regions of both hydrophobic and hydrophilic character. A direct result of these complex variations in wool morphology is that kinetics of diffusion and polarity of the reactants may be at least as important as inherent reactivity.

生词与词组

1.husbandry [ˈhʌzbəndri]n. 畜牧；牧业

2.pastoral [ˈpɑ:st(ə)rəl]n. 放牧者 adj. 牧人的

3.shearing [ˈʃɪərɪŋ]n. 剪毛；剪短；剪羊毛

4.clipper [ˈklɪpə(r)]n. 剪具；剪子；剪取人

5.blade [bleɪd]n. 刀片；刀口；剑；刀

6.flock [flɒk]n. 羊群

7.fleece [fli:s]n. 羊毛；羊毛状物；盖满

8.woolly sheepskin 绵羊毛皮

9.slaughtered sheep 屠宰的羊

10.lamb [læm]*n*. 羔羊；小羊

11.pelt [pelt]*n*. 毛皮；生皮

12.full fleece 全毛套；全绒毛

13.crutching [ˈkrʌtʃɪŋ]*n*. 粪污碎毛；腿臀毛

14.hindquarter [ˈhaɪndˌkwɔː(r)tə(r)]*n*. 粪污碎毛

15.clip preparation 套毛整理

16.contamination [kənˌtæmɪˈneɪʃ(ə)n]*n*. 污染；污物；污秽

17.belly wool 腹毛

18.discoloured crutch 变色腋下毛

19.discriminate [dɪˈskrɪmɪneɪt]*v*. 区别；辨别

20.tenderness [ˈtendənəs]*n*. 柔软

21.wool scouring 洗毛

22.carbonising [ˈkaːbəˌnəɪzɪy]*n*. 碳化

23.greasy [ˈgriːsi]*adj*. 多油的；多脂的；带油污的；油腻的

24.impurity [ɪmˈpjʊərəti]*n*. 杂质

25.topmaking 毛条加工

26.minimise entanglement 最小化纠缠

27.contaminant [kənˈtæmɪnənt]*n*. 污染物；致污物

28.luxury handle 豪华手感；华丽手感

29.morphology [mɔː(r)ˈfɒlədʒi]*n*. 形态；微观形貌；组织

30.breed [briːd]*v*. 饲养

31.domesticated [dəˈmestɪˌkeɪtɪd]*adj*. 驯养的

32.cashmere [ˈkæʃˌmɪə(r)]*n*. 山羊绒；开司米

33.Angora [æŋˈgɔːrə]*n*. 安哥拉山羊

34.camel [ˈkæm(ə)l]*n*. 骆驼

35.vicuna [vɪˈkjuːnə]*n*. 骆马

36.alpaca [ælˈpækə]*n*. 羊驼

37.Angora rabbit 安哥拉兔

38.yak [jæk]*n*. 牦牛

39.keratin [ˈkerətɪn]*n*. 角蛋白

40.silk [sɪlk]*n*. 丝

41.feather [ˈfeðə(r)]*n*. 羽毛

42.macromolecular [ˌmaːkrəʊˈmɒliˌkjuːlə]*adj*. 大分子的；高分子的

43.fibril-reinforced matrix material 纤维增强基体材料

44.fibril ['faɪbrɪl]*n.* 原纤；纤丝

45.diffusion [dɪ'fjuːʒ(ə)n]*n.* 扩散

46.reactant chemical 化学反应物

47.immersion solvent 浸渍溶剂

48.hydrophobic epicuticle 疏水性表皮

49.micro-heterogeneous 细观非均质

50.kinetics [kaɪ'netɪks]*n.* 动力学

51.polarity [pəʊ'lærəti]*n.* 极性

52.inherent reactivity 固有活性

译文

第7课　羊毛

养羊是欧洲、美洲和亚洲大部分地区牧民的一项重要活动。尽管数百年来羊毛一直是纺织工业的重要原料,但现在它的主要目的是生产肉类。

1.收获

大多数羊毛是用电动手推剪剪羊毛。对于可能发生的恶劣天气,刀片剪羊毛仍然是首选,因其可留下大约10毫米的羊毛作为防寒保护。屠宰绵羊或羔羊的羊皮经过化学处理以削弱纤维根,以便羊毛从毛皮上拉下来。

羊毛分为全羊毛(一年生长,如图7.1)、二次剪毛和早剪毛(不到一年生长)和粪污碎毛(剪下的羊后腿及臀部的毛)。

图 7.1　全绒羊毛

2. 套毛整理

在剪毛环节,每块羊毛都需检查,以去除羊毛中的杂色毛、异长毛和污渍毛,并将其分别收集。疵点毛分为肚皮毛、变色腋下毛、污渍毛等。羊毛中的植物质、黏结毡片也应分离去除。细羊毛生产线需区分羊毛的纤维直径、长度、强度、颜色和植物杂质含量等参数。变色和羊毛柔软度是地毯用羊毛的分选原则。

3. 洗毛、碳化

原毛或粘有"油腻"的毛,其类型取决于羊的品种、饲养羊的地区和饲养方法。洗毛的作用是:

(1)清除毛中的污染物。

(2)确保羊毛的物理和化学条件适合预期的加工路线(如最大程度地减少缠结使毛条保持短纤维结构)。

术语"洗毛"是从原毛中去除污染物的过程。因此,它包含清洁羊毛的所有过程,包括水清洁以外的溶剂清洗过程和固体作为载体去除污染物的过程。显然,洗毛是羊毛加工过程中重要环节。

4. 纤维直径

测量纤维直径能更好解释细羊毛和粗羊毛之间的差异。对于细美利奴羊毛,批次内的平均细度在 18-21 微米之间,且该范围小直径的纤维有更高价格。当然,平均直径与纺纱极限和华丽手感密切相关,纤维细度每细 1 微米可获得更高价值。

图 7.2   安哥拉山羊

5. 纤维形态

"wool"是对各种驯养绵羊毛的通用描述,尽管它也常用作所有动物毛的通称,特别是动物细毛,即羊绒、安哥拉山羊毛(安哥拉山羊,如图 7.2)、骆驼毛、骆马毛和羊驼毛、安哥拉兔毛以及牦牛毛。人发的形态和成分也与羊毛非常相似。羊毛含有 α- 角蛋白,而丝绸和羽毛则由 β - 角蛋白组成。从大分子的角度来看,羊毛是一种复合纤维,即一种原纤维增强的基质材料,原纤维和基质均由多肽组成,二者在物理和化学形态上相互连接。从形态学的角度来看,羊毛纤维是一种高度复杂的纳米复合材料,具有清晰的层次结构,其具有极高的自组织性。

6. 羊毛化学性

在所有羊毛类型中,纤维直径范围为 10–80 毫米。浸渍溶剂等(最常见的是水)化学反应物的扩散可能很慢。除了纤维直径的变化,溶剂扩散也受到羊毛鳞细胞表面的疏水性角质层影响。

每个细胞内的蛋白质组织都存在差异,导致羊毛具有疏水性和亲水性的微异质区域。羊毛形态的这些复杂变化的直接结果是,反应物的扩散动力学和极性差,与其固有反应性一样重要。

Lesson 8　Mohair

Mohair is largely produced in South Africa and the United States of America but also in Turkey, Argentina, Australia and New Zealand. South Africa presently accounts for approximately 60% of the world production of mohair.

General characteristics of mohair

Mohair is characterised by excellent lustre, durability, elasticity, resilience, abrasion resistance, draping, moisture and perspiration absorption and release, insulation comfort and pleasing handle, and by low flammability, felting and pilling.

Its good insulation makes mohair fabrics light-weight and warm in winter and comfortably cool in summer, which is also a function of the fabric and garment construction. Although mohair has proved extremely popular in many applications, it has some limitations in certain apparel applications, because of its coarseness relative to other types of apparel fibres, such as cotton. Its outstanding properties, such as resilience and durability, also make it particularly suitable for household textiles, such as upholstery fabrics, curtains and carpets.

Mohair's lustre, smoothness, low friction, low felting and certain other properties are all related to its surface scale structure, the scales generally being thin and relatively long. Mohair shares many of the outstanding properties of other animal fibres.

Flammability

Mohair has low flammability, in common with other animal fibres such as wool. When exposed to a naked flame, it burns at a low temperature and tends to shrink. The flame produces a bead-like ash, but the fibre will stop burning almost as soon as it is taken away from the flame.

Durability

Because mohair's structure is pliable, it can be bent and twisted repeatedly without damage to the fibre, making it one of the world's most durable animal fibres.

Elasticity

Mohair is very elastic. A typical mohair fibre can be stretched to 130% of its normal length and will still spring back into shape. Because of the fibre's resilience, mohair garments resist wrinkling, stretching, and bagging during wear.

Moisture absorption

Mohair can absorb moisture from the atmosphere readily (up to 30% without feeling wet). Because mohair dries slowly the danger of getting a chill is reduced.

Setting

Mohair may be set to retain extension or deformation more readily than most other animal fibres. The fibre's setting ability is capitalised on in the manufacture of curled-pile rugs.

Lustre

Mohair's well-known lustre is caused by its closed (unpronounced) scale formation and can be preserved or even enhanced by careful processing and dyeing.

Dyeing

It is possible to dye mohair brilliant colours that resist time, the elements, and hard wear. Due to this property mohair gets the name "the Diamond Fibre".

Soiling resistance

Because of its smoothness and other characteristics, mohair generally exhibits good soil resistance and desoiling.

Felting

Mohair has a very low tendency to felt.

Light weight

Mohair blends well with wool and can produce smooth yarns, enabling fabrics to be produced which are noted for coolness, such as lightweight summer fabrics. It is unsurpassed in tropical suitings, largely because it combines coolness with durability; the material is also effective when made into linings because of its good moisture absorption and drape characteristics.

Length

Prized as a textile fibre because of its length, mohair fibre averages about 300mm for a full year's growth (i.e., 25mm per month). For example, exceptionally long fibres (up to 300mm) are used to make women's switches, doll's hair and theatrical wigs.

生词与词组

1.durability [ˌdjʊərəˈbɪlətɪ]*n.* 耐久性；耐用性；耐候性

2.resilience [rɪˈzɪliəns]*n.* 弹性；恢复能力；回弹性

3.abrasion resistance 耐磨性；抗磨损

4.draping [ˈdreɪpɪŋ]*n.* 悬挂；悬垂状态

5.perspiration absorption 排汗

6.insulation comfort 保温舒适性；保暖性

7.flammability [ˌflæməˈbɪlətɪ]*n.* 易燃性；可燃性

8.felting [ˈfeltɪŋ]*n.* 毡；毡合

9.pilling [ˈpɪlɪŋ]*n.* 起球

10.upholstery [ʌpˈhəʊlst(ə)ri]*n.* 家具装饰品；室内装饰品

11.curtain [ˈkɜ:(r)t(ə)n]*n.* 窗帘；门帘

12.smoothness [ˈsmu:ðnəs]*n.* 平滑；柔滑；光滑；丝般光滑

13.naked flame 明火

14.bead-like ash 珠状灰烬

15.pliable [ˈplaɪəb(ə)l]*adj.* 易受影响的；易弯的

16.bent [bent]*adj.* 弯曲的；弯的

17.wrinkling [ˈrɪŋklɪŋ]*n.* 褶皱；起皱；皱曲

18.bagging [ˈbægɪŋ]*n.* 使膨胀；拱胀

19.chill [tʃɪl]*n.* 寒冷；着凉 *v.* 冷藏；冷却；使变冷 *adj.* 寒冷的

20.curled-pile rug 卷毛地毯；卷绒地毯

21.unpronounced [ˌʌnprəˈnaʊnst]*adj.* 未发育的

22.dyeing [ˈdaɪɪŋ]*n.* 染色 *adj.* 染色的

23.lining [ˈlaɪnɪŋ]*n.* 衬料；衬里

24.theatrical wig 戏剧假发

译文

第8课　马海毛

马海毛主要产自南非和美国，土耳其、阿根廷、澳大利亚和新西兰也有生产。目前，南非马海毛产量约占世界的60%。

马海毛的一般特征：

马海毛具有优异的光泽、耐用性、弹性、回弹性、耐磨性、悬垂性、吸湿排汗、保暖性、手感，以及低可燃性、毡化和起球性。

良好的隔热性使马海毛面料轻薄、冬暖夏凉，这也是面料和服装的结构功能。尽管马海毛在许多应用中非常受欢迎，但在某些服装应用中具有一些局限性，因为它相对于其他类型的纤维（如棉）更粗糙。出色的弹性和耐用性，使其特别适用于家用纺织品，如室内装饰织物、窗帘和地毯。

马海毛的光泽、光滑度、低摩擦、低毡化以及其他特性都与其表面鳞片结构有关，其鳞片通常很薄且相对较长。马海毛具有许多其他动物纤维的突出特性。

1. 燃烧性

与羊毛等其他动物纤维一样，马海毛可燃性较低。当暴露在明火中时，它会在低温下燃烧且趋于收缩。燃烧产生珠状灰烬，但纤维离开火焰就会停止燃烧。

2. 耐用性

由于马海毛的结构柔韧，可反复弯曲和扭曲而不会损坏纤维，使其成为世界上最耐用的动物纤维之一。

3. 弹性

马海毛很有弹性。典型的马海毛纤维可以拉伸至其正常长度的130%，并仍会回复原样。因马海毛纤维的回弹性好，其服装在穿着过程不易起皱，抗拉伸，抗拱胀。

4. 吸湿性

马海毛极易从大气中吸收水分(其含水率高达 30%,无湿感)。因马海毛变干速度缓慢,降低了人受凉的风险。

5. 定型

与大多数其他动物纤维相比,马海毛定型后更易保持伸展或变形。在卷毛地毯的加工中,充分体现了这种纤维的定型能力。

6. 光泽

马海毛的美丽光泽是因其封闭(未发育)的鳞片构成引起的,可通过精加工和染色来保留,甚至增强其光泽。

7. 染色性

马海毛可染鲜艳的颜色,可耐时间、化学元素和长期穿着的影响。"钻石纤维"这个名字由此而来。

8. 耐污性

由于其纤维表面光滑及其他特性,马海毛展现出良好的防污和去污能力。

9. 毡化

马海毛毡化趋势非常低。

10. 轻量

马海毛与羊毛混纺可生产出光滑的纱线,可织制凉爽的面料,如轻质的夏季面料。马海毛面料凉爽,耐用性强,是热带套装面料的首选;因其良好的吸湿性和悬垂性,该面料也可制成优异的衬料。

11. 长度

马海毛因纤维长度适宜做纺织纤维,全年生长的纤维平均长度约为 300 毫米(即每月生长 25 毫米)。例如,特长的纤维(长达 300 毫米)可用于制作女性的发辫、娃娃的头发和戏剧假发。

Lesson 9　Cashmere

The cashmere goat and its fibre takes its name from Kashmir. At present little fibre is obtained from that area and cashmere is now principally produced in northern China and Afghanistan. Smaller quantities are also produced in Iran, Australia and New Zealand.

The height of these goats is between 60cm and 80cm. Their average life span is about 7 years. The fleece is open, with long coarse outer hair and under hair or down. Each goat produces between 100g and 160g of usable down per year. The down has an average length of 35mm–50mm. The fine down enables these goats to withstand the extreme winter cold of their original habitat, the plateaux of Central Asia. They protect themselves from overheating in the summer by shedding their down in the spring.

Fibre production, harvesting and characteristics

Of the world's production of 9,000–10,000 tonnes, 50%–60% comes from China, 20%–30% from Mongolia, and the balance from Iran and Afghanistan. Production in China has fallen by 10% during the past few years due to severe winter conditions. China is also attempting to control goat numbers because of overgrazing which causes desertification.

In China, cashmere is harvested by combing during the three-to-six-week spring period when the goats are moulting or by collecting the moulted fibres from the ground and bushes. In Iran, Afghanistan, Australia and New Zealand the fleece is usually shorn.

The hair is sorted by hand for grades and colours (white, grey and brown). This is done quickly and requires considerable expertise, and reduces the amount of guard hair. After sorting, the different piles of hair are "willowed", which entails putting the fibres through a simple revolving machine to shake out much of the dust and grit. After sorting and willowing the fibres are scoured.

The quality of the dehaired fibre is assessed by the diameter, colour and length and the coarse hair content. Diameters are within the range of 14mm–19mm and the fibre lengths measure from 150mm to 450mm. Chinese cashmere (see Fig. 9.1)is considered to be of the best quality, has a fibre diameter of 14mm–16mm, and is predominantly white.

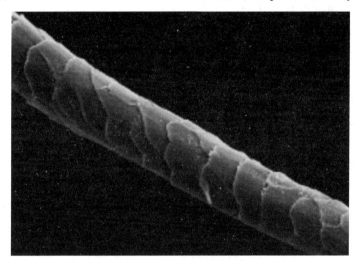

Figure 9.1   Cashmere fibres

生词与词组

1.life span 寿命

2.plateau [ˈplætəʊ]n. 高原；高地

3.overgrazing [ˈəʊvəˈgreizɪŋ]n. 过度放牧；超载放牧

4.desertification [dɪˌzɜ:(r)tɪfɪˈkeɪʃ(ə)n]n. 沙漠化；荒漠化

5.moulting [məʊltɪŋ]n. 脱皮；蜕皮

6.entail [ɪnˈteɪl]v. 使……成为必须；需要；引起

7.guard hair 针毛；外层粗毛；粗硬毛

8.grit [grɪt]*n.* 沙砾；粗砂；沙粒

9.sorting and willowing 分拣和清理

译文

第 9 课　山羊绒

克什米尔山羊及其纤维的名字来自克什米尔。目前,该地区(克什米尔)出产的纤维很少,现在中国北部和阿富汗是山羊绒主产地。伊朗、澳大利亚和新西兰也有少量出产。

这些山羊的身高为 60~80 厘米。它们的平均寿命约为 7 年。羊毛外露,包括长而粗的外毛、底毛或底绒。每只山羊每年生产 100~160 克绒毛。绒毛的平均长度为 35~50 毫米。细细的绒毛使这些山羊能够抵御其原始栖息地——中亚高原的严寒。在春季山羊脱掉绒毛,使其抵御夏季的炎热。

纤维生产、收获和特性:

全世界山羊绒总产量为 9000~10000 吨,50%~60% 产量来自中国,20%~30% 产量来自蒙古国,其余来自伊朗和阿富汗。前几年,受严寒影响,中国的羊绒产量降了 10%。因过度放牧导致草原荒漠化,中国正努力控制山羊的数量。

在中国,羊绒是在山羊换绒的 3~6 周(春季)时间经梳理获得,或从地面、灌木丛中收集脱落的纤维。在伊朗、阿富汗、澳大利亚和新西兰,通常是剪取羊毛。

毛发需按等级和颜色(白色、灰色和棕色)手工分类。手工分类完成得很快,需要大量专门技能才可完成,以减少针毛量。经分拣,挑出粘结成鞭的毛发,其需放入旋转机器以去除大量灰尘和沙砾。分选和清理后,纤维需进行洗涤。

纤维直径、颜色和长度以及粗毛含量是评价脱毛纤维质量的指标。山羊毛纤维的直径范围为 14~19 毫米,其纤维长度范围为 150~450 毫米。中国羊绒(见图 9.1)质量最好,纤维直径为 14~16 毫米,以白色为主。

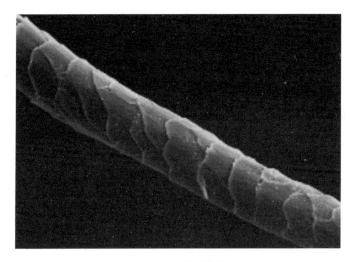

图 9.1   山羊绒纤维

Lesson 10　Other animal hair fibres

The fine undercoat (down) fibres, of two-coated animals, which are generally shed during spring, are the most valuable from a textile perspective, because of their combination of fineness, softness, lightness and good thermal insulation, and need to be separated from the undesirable coarse guard hair, either by hand or mechanically, a process called dehairing. The finer the fibre and the lower the percentage of coarse fibres remaining in the fine component after dehairing, the better the textile quality and value of the fibre. In this respect, successful dehairing is largely dependent upon the differences between the two components in terms of fibre diameter/rigidity/linear density, friction, length and inter-fibre cohesion.

Essentially, the main speciality animal hair fibres covered can be grouped into the following two main families or groups:

Camel
  – Alpaca
  – Bactrian camel
  – Llama

Bovine
  – Musk-ox
  – Yak

Most luxury or speciality animal fibres tend to be finer, less crimped and smoother than wool, their cuticular scales also being less pronounced (flatter or thinner), typically 0.4μm thick or even thinner, compared to those of wool which is generally 0.6μm and thicker. The scales are also more widely spaced. Where crimp is present, they are generally not as pronounced as that of fine wools and in some cases is better described as a curling. Fibre cohesion and friction are consequently lower than those of wool, requiring special conditions, or blending with other fibres, such as wool, for acceptable mechanical processing performance and yarn quality. After dehairing, the down fibres are generally processed into yarn following either the worsted route for the longer fibres or the woollen route for the shorter fibres. Mechanical processing, although mostly done on machinery similar to that used for wool, needs to be adapted and optimised to suit the specific requirements of each of these fibres. Chemically, these fibres belong to the same protein (keratin) family of fibres as wool, although their fine morphological and surface structures do differ.

These hair fibres are mostly processed, dyed and finished using similar machinery to that for wool, but the settings and the conditions (e.g., dyeing and finishing recipes, etc.) are adjusted to suit the specific characteristics, notably length and fineness, of each fibre. Because of the smooth and medullated nature of many of these speciality fibres, which differs from that of most apparel wools, dyeing recipes need to be adjusted from those used for wool, if a specific dye shade is to be achieved. Great care is also taken in practice to select the dyeing and finishing conditions, such as treatment time, temperature and pH, so that the desirable characteristics of the fibres, for example lustre and softness, are not deleteriously affected.

生词与词组

1.undercoat [ˈʌndə(r)ˌkəʊt]n. 短毛；(动物长毛下面的) 浓密绒毛

2.combination of fineness 细度组合

3.thermal insulation 隔热

4.dehairing [diːˈheərɪŋ]n. 脱毛；拔毛

5.rigidity [rɪˈdʒɪdəti]n. 僵直；刚度；刚性；僵硬

6.cohesion [kəʊˈhiːʒ(ə)n]n. 内聚力；凝聚；结合力

7.Bactrian camel 骆驼；双峰驼

8.llama [ˈlɑːmə]*n*. 美洲驼；无峰驼

9.bovine [ˈbəʊvaɪn]*n*. 牛科动物

10.musk-ox[mʌskɒks]*n*. 麝牛

11.crimped [krɪmpt]*adj*. 卷曲的

12.cuticular scale 角质鳞片

13.flatter [ˈflætə(r)]*adj*. 平的；扁的

14.curling [ˈkɜː(r)lɪŋ]*n*. 卷曲；卷边；卷缩

15.blending with 混合

16.worsted route 精纺工艺

17.woollen route 粗纺工艺

18.recipe [ˈresəpi]*n*. 食谱；处方

19.deleterious [ˌdeləˈtɪəriəs]*adj*. 有害的；有毒的；造成损坏的；导致伤害的

译文

第 10 课　其他动物毛纤维

从纺织品的角度来看,动物在春季脱落的细绒毛是最具纺用价值的,因其集细度、柔软度、轻盈度和隔热性于一体,需用手或机械从中分离粗毛,这一过程称为"捡毛"。纤维越细,脱毛后残留在细纤维中的粗纤维的百分比越低,制成纺织品的质量和价值就越好。这方面,脱毛很大程度上取决于两种成分在纤维直径/刚性/线密度、摩擦性、长度以及纤维间内聚力方面的差异。

基本上,特种动物毛纤维主要分为以下两组:骆驼(羊驼、双峰驼、美洲驼)、牛(麝牛、牦牛)。

大多数特种动物纤维比羊毛更细,卷曲更少,更光滑,它们的表皮鳞片也不太明显(更平或更薄),其鳞片厚度通常为 0.4 微米或更薄,而羊毛鳞片厚度通常为 0.6 微米或更厚。鳞片的间距也更宽。屈曲时,特种动物纤维通常不如细羊毛明显,在某些情况下,最好将其称为卷曲。特种动物纤维的内聚力和摩擦力低于羊毛,为获得足够的机械性能及良好的纱线质量,其需要进行特殊加工,或与其他纤维(如羊毛)混纺。脱毛后,较长绒毛纤维可进行精梳纺纱,较短绒毛纤维可进行粗梳纺纱。特种动物毛纤维的加工机械类似于羊毛加工机械,但仍需要进行调整、优化,以适应每种纤维的具体要求。化学方面,这些纤维与羊毛细观形态和表面结构确实不同,但同属蛋白质(角蛋白)纤维。

这些毛发纤维大多使用与羊毛相似的机器进行加工、染色和整理,但需调整配置和条件(如染整配方等)以适应每种纤维的差异(尤其是长度、细度)。

由于这些特种纤维光滑且有髓腔,与大多数服用羊毛不同,如要获得特定的染色色调,则需依据羊毛的染色配方进行调整。在实践生产中,需谨慎选择染色和整理条件(如处理时间、温度和 pH 值),以使理想纤维特性,如纤维光泽和柔软度不受有害的影响。

Lesson 11　Silk fibres

　　Silk is one of the oldest fibres known to man. Silk has been used as a textile fibre for over 4,000 years. Silk is an animal fibre produced by certain insects to build their cocoons and webs. Over the centuries, silk has been a highly valued textile fibre. Despite facing keen competition from man-made fibres, silk has maintained its supremacy in the production of luxury apparel and specialized goods of the highest quality.

　　Silk fibres are remarkable materials displaying unusual mechanical properties: strong, extensible, and mechanically compressible. Silk is rightly called the queen of textiles for its lustre, sensuousness and glamour. Silk's natural beauty and properties of comfort in warm weather and warmth during colder months have made it useful in high-fashion clothing.

1.Classification

　　Sericulture is the rearing of silkworms for the production of raw silk. The major activities of sericulture comprise food-plant cultivation to feed the silkworms which spin silk cocoons and reeling the cocoons for unwinding the silk filament for value-added benefits such as processing and weaving. There are five major types of silk of commercial importance, obtained from different species of silkworms which in turn feed on a number of food plants.

Figure 11.1　Silk reeling

Except mulberry, other varieties of silks are generally termed non-mulberry silks. India has the unique distinction of producing all of these commercial varieties of silk. The bulk of the commercial silk produced in the world comes from this variety and the silk is referred to as mulberry silk. Mulberry silk comes from the silkworm.

2.Silk reeling

Silk reeling (see Fig. 11.1)is the process of unwinding the silk filaments from the cocoons and the process by which a number of cocoonbaves are reeled together to produce a single thread. This is achieved by unwinding filaments collectively from a group of cooked cocoons at one end in a warm water bath and winding the resultant thread onto a fast moving reel. Raw silk reeling may be classified by the direct reeling method on a standard sized reel, the indirect method of reeling on small reels, and the transfer of reeled silk from small reels onto standard sized reels on a re-reeling machine.

3.Silk throwing

The first process of manufacture through which raw reeled silk must pass corresponds in some ways to the carding, combing and spinning in cotton and wool. In silk manufacturing it is called throwing. It is a process in which the silk strands are twisted together with other silk strands to form a thicker, stronger, multi-threaded yarn. Throwing produces a wide variety of yarns that differ according to the number of strands and the amount and direction of the twist imparted.

When the silk arrives at the throwing mills it is usually in the form of skeins. Throwing naturally does not include the common processes of carding and combing, for the reason that the reeled silk is already in the form of thread. Throwing is essentially a process of cleaning, doubling and twisting the single fibres as they come from the filatures.

4.Microstructure and appearance

Silk fibres spun out from silkworm cocoons consists of fibroin in the inner layer and sericin in the outer layer. Each raw silk thread has a lengthwise striation, consisting of two fibroin filaments of each embedded in sericin. The chemical compositions are, in general, silk fibroin of 75%–83%, sericin of 17%–25%, waxes of about 1.5%, and others of about 1.0% by weight. Silk fibres are biodegradable and highly crystalline with well-aligned structure. It is known that they also have higher tensile strength than glass fibre or synthetic organic fibres, good elasticity and excellent resilience. Silk fibre is normally stable up to 140℃ and the thermal decomposition temperature is higher than 150 ℃ . The densities of silk fibres are in the range of $1,320kg/m^3$–$1,400kg/m^3$ with sericin and $1,300kg/m^3$–$1,380kg/m^3$ without sericin.

5.Amino acid composition

The amino acid composition varies in different varieties of silk. Three major amino acids including serine, glycine and alanine may be found in mulberry and non-mulberry varieties.

6.Properties of silk

Silk fibers are remarkable materials displaying unusual mechanical properties. They also display interesting thermal and electromagnetic responses, particularly in the ultraviolet (UV) range and form crystalline phases related to processing. The mechanical properties of silk fibers are a direct result of the size and orientation of the crystalline domains. Other properties of silk such as good thermal stability, optical responses, dynamic mechanical behavior and time dependent responses have all been used in number of applications in various fields.

7.Applications of silk

Silk is one of the most beautiful fabrics available, with a long and colourful history and changing applications in the world today. Be it for gowns, medical use, home decor and more, the uses of silk constitute a wide and varied topic.

生词与词组

1.cocoon [kəˈkuːn]n. (蚕) 茧
2.mulberry [ˈmʌlb(ə)ri]n. 桑树；桑；桑属；桑葚
3.moth [mɒθ]n. 飞蛾；蛾类昆虫
4.keen competition 激烈竞争

5.glamour [ˈɡlæmə(r)]n. 魅力

6.sericulture [ˈserɪˌkʌltʃə]n. 蚕业；桑蚕业

7.rear [rɪə(r)]v. 饲养；养育

8.reeling [ri:lɪŋ]n. 缫丝；摇丝；摇纱；卷取

9.unwinding [ʌnˈwaɪndɪŋ]n. 退绕；松卷

10.carding [ˈkɑ:(r)dɪŋ]n. 梳型；梳棉；分梳；粗纺

11.combing [ˈkəʊmɪŋ]n. 精梳

12.throwing [ˈθrəʊɪŋ]n. 制丝；丝织准备；摇丝；打线

13.multi-threaded [mʌlˈtɪθɪˈedɪd]adj.多股的；多线的

14.filature [ˈfɪlətʃə]n. 缫丝厂；缫丝机；缫丝；制丝厂

15.skein [skeɪn]n.(一) 绞；绞纱；绞丝；缕丝

16.sericin [ˈseərɪsɪn]n. 丝胶蛋白；丝胶

17.lengthwise striation 纵向条纹

18.well-aligned [ˈweləlˈaɪnd]adj. 排列整齐的

19.amino acid 氨基酸

20.serine [ˈserɪn]n. 丝氨酸

21.glycine [ˈɡlaɪsi:n]n. 甘氨酸

22.alanine [ˈæləni:n]n. 丙氨酸

23.electromagnetic [ɪˌlektrəʊmæɡˈnetɪk]adj. 电磁的

24.ultraviolet [ˌʌltrəˈvaɪələt]adj. 紫外线的 n. 紫外线

译文

第 11 课　蚕丝

丝绸是人类已知最古老的纤维之一。丝绸被用作纺织纤维已有
4000 多年的历史。丝绸是由某种昆虫产的一种动物纤维,其用于构建
昆虫的茧。几个世纪以来,丝绸一直是极具价值的纺织纤维。尽管面对
人造纤维的激烈竞争,丝绸在生产高档服装和最高品质的专用商品方面
仍保持领先地位。

丝纤维具有不同寻常的机械性能:坚固、可延展和易压缩。丝绸因
其光泽华丽被誉为"纺织品女皇"。丝绸具有华丽外观和冬暖夏凉特性,
使其成为高级时装首选。

1. 分类

蚕业是指养蚕以生产生丝。蚕业的主要活动是种植食用植物以喂养蚕,得蚕茧,蚕茧经抽丝、缫丝以获得纺用丝线,以实现加工和织造等。具有商业价值的蚕丝有五种,其来自以不同植物为食的不同品种的蚕。

除桑蚕丝外,其他品种的蚕丝一般称为非桑蚕丝。在所有商业品种的蚕丝生产方面,印度具有独特的优势。世界上大部分商业丝绸都来自桑蚕丝。桑蚕丝来自蚕。

2. 缫丝

缫丝(如图 11.1)是将丝从茧中抽取的过程,也是将多根茧丝缠绕在一起产生单线的过程。抽丝是从温水浴的熟茧端部抽取丝,并将丝快速缠绕到卷轴上。生丝缫丝可分为标准卷筒直卷法,小卷筒间接卷取法,小卷筒卷装再导成标准卷筒。

图 11.1　缫丝

3. 捻丝

生缫丝必经的第一步加工,即是与棉和羊毛类似的开松、梳理和纺制过程。在丝加工中,它被称为捻丝。它是将多股丝线捻合在一起,形成更粗、更结实的多股纱线的过程。捻丝生产的纱线种类繁多,其依据股线的数量、所施加的捻度和方向的不同而异。

丝通常以绞纱的形式到达捻丝厂。因缫丝环节丝已是线的形式,所以其捻丝过程不包括梳理和精梳工序。捻丝是清理、并股和捻合单丝的过程。

4. 显微结构和外形

蚕茧的丝纤维由内层的丝蛋白和外层的丝胶组成。每根生丝都有一条纵向条纹,其由两条丝素蛋白组成,每条丝素蛋白都嵌入丝胶中。丝化学成分由重量占 75%-83% 的丝素蛋白、17%-25% 的丝胶、约 1.5% 的蜡和约 1.0% 的其他物质组成。丝纤维可生物降解,结晶度高,结构排列整齐。丝还具有比玻璃纤维或合成纤维更高的拉伸强度、良好的弹性和回弹性。在 140℃ 以下,蚕丝纤维较稳定,温度大于 150℃ 易热分解。丝纤维(含丝胶)的密度为 1320-1400 千克 / 立方米,丝纤维(不含丝胶)的密度为 1300-1380 千克 / 立方米。

5. 氨基酸成分

不同品种的蚕丝的氨基酸组成存在差异。桑蚕丝和非桑蚕丝含有丝氨酸、甘氨酸和丙氨酸 3 种氨基酸。

6. 丝的特性

丝纤维是一种具有不同寻常机械性能的材料。它们还展示出热和电磁响应,特别在紫外线(UV)条件下可形成与加工相关的结晶相。丝纤维的机械性能是与结晶域的大小和取向息息相关的。丝绸的其他特性,如良好的热稳定性、光学响应、动态机械行为和时间响应,都已用于各个领域。

7. 丝的应用

丝绸是当今世界上最华丽的面料之一,其有着悠久而多彩的历史和多样的应用。无论是长袍、医疗用途,还是家居装饰等,丝绸的应用主题已多样化。

Lesson 12   Feathers and down

1.Feathers and down

Feathers and down are two of nature's most marvelous products. Together they form the bird's plumage, but they are very different in structure. A feather has a two-dimensional structure, a quill from which barbules extend in two opposite directions like vanes and a compact and flat tip. Down has no quill but a small core, from which small clusters of barbs (with barbules and nodes) extend in three dimensions. It would be erroneous to assume that down is a small feather or would eventually develop into a feather. Feathers form the outer, protective coat of the animal's body. They have structural functions, best seen in the strong wing and tail feathers. They are highly resilient. Down, only found in waterfowl, provides the necessary warmth insulation for the bird. The barbs are fluffy-elastic and at least as resilient as feathers. Thanks to its structure, down can trap a large volume of air, resulting in an almost unsurpassed heat insulation capability with respect to weight.

2.Test standards

In order to write, read, compare and understand the quality or performance data of down and feathers in a universal language, the International Down and Feather Bureau (IDFB), together with international and national standardization organizations, has developed methods and published standards for both testing and characterizing down and feather material and products. When applying any of these methods, it is important that the procedures are followed in their entirety. The characterization and testing of all materials, natural or synthetic, are determined by their physical structure and chemical composition.

3.The physical structure of feathers and down

Figure 12.1　Micrograph of a feather

Figure 12.2　Micrograph of a down cluster

The barbs are not symmetrical (see Fig. 12.1). Furthermore, it is clear that an individual down cluster is not at all as symmetric as an individual snowflake, with which down is sometimes compared (see Fig. 12.2).

生词与词组

1.feather [ˈfeðə(r)]*n.* 羽毛；翎羽

2.down [daʊn]*n.* 绒毛；软毛；汗毛

3.marvelous [ˈmɑ:(r)vələs]*adj.* 奇异的；非凡的

4.plumage [ˈplu:mɪdʒ]*n.* 鸟类羽毛；羽毛；羽衣；羽被

5.quill [kwɪl]*n.* 羽毛管；羽茎

6.flat tip 平翼尖（梢）

7.barbule [ˈbɑ:bju:l]*n.* 羽小枝

8.vane [veɪn]*n.* 叶轮；叶；羽片；短毛

9.clusters of barb 簇绒的羽枝

10.node [nəʊd]*n.* 节；结

11.resilient [rɪˈzɪliənt]*adj.* 可迅速恢复的；回弹的

12.waterfowl [ˈwɔ:tə(r)ˌfaʊl]*n.* 水鸟；水禽

13.wing [wɪŋ]*n.* 翼；翅

14.tail [teɪl]*n.* 尾；尾翼

15.snowflake [ˈsnəʊˌfleɪk]*n.* 雪花

16.insulation capability 保暖性能；隔热性能

译文

第 12 课　羽毛和羽绒

1. 羽毛和羽绒

羽毛和羽绒是自然界最神奇的两种产品。它们均来自鸟的羽毛,但它们的结构却大不相同。羽毛具有二维结构,小枝从羽毛管向两个方向延伸(如羽片、翼尖、羽枝),尖端紧凑而扁平。羽绒没有羽茎,只有一个核心,簇绒(带有小枝和节点)由其向三维延伸。认为羽绒是一根小羽毛或最终会发展成羽毛是错误的观点。羽毛是动物身体的外保护层。它们具有结构功能,在强壮的翅膀和尾羽中最为明显。它们具有很强的弹性。只有水禽长有羽绒,羽绒为鸟类提供必要的保暖。簇绒蓬松,和羽毛一样有弹性。因独特结构,羽绒可捕获大量空气,使其具有无与伦比的隔热和超轻性能。

2. 测试标准

为了便于使用通用语言书写、阅读、比较和理解羽绒羽毛的质量或性能数据,国际羽绒羽毛局(IDFB)协同国内外标准化组织制定并发布了羽绒羽毛材料、产品测试和表征方法的标准。在使用任何一种方法时,执行者必须完全遵循标准流程。所有材料(天然或合成材料)的特性和测试由它们的物理结构和化学成分决定。

3. 羽绒羽毛的物理结构

图 12.1　一根羽毛的显微照片

羽枝不对称(见图 12.1)。此外,有时会将羽绒与雪花进行比较,显然单个羽绒簇根本不像单个雪花那样对称(见图 12.2)。

图 12.2   簇绒的显微照片

Lesson 13   Industrial cellulose

1.Pulping methods

Chemical and mechanical pulping of wood raw materials dominate the paper and board industry worldwide, although other fibre materials called nonwood are used in certain countries. The most common chemical pulping process is the kraft pulping process, which is an alkaline process utilising sodium hydroxide and sodium sulphide as active delignification chemicals. The kraft pulping method is able to process a variety of fibre raw materials including both softwood and hardwood raw materials.

2.Dissolving pulp processes

The pulping process has to dissolve the main bulk of the lignin and modify the residual lignin for successful bleaching. Bleaching then has to remove this residual and increase the pulp brightness and cleanliness. The final result should be a technical cellulose as free as possible from lignin and hemicelluloses as well as extractables. The α-cellulose content of the final dissolving pulp may normally vary between 90% and 96%, dependent upon the pulping and bleaching processes.

3.The viscose process

The first patent on the viscose process was granted to Cross and Bevan in England in 1893. By 1908 the fibre spun from viscose dope had been accepted as a key component of the burgeoning textile industry. Viscose (or rayon) still enjoys the unique position of being the most versatile of all artificial fibres. This has resulted from an ability to engineer the fibre chemically and structurally in ways that take advantage of the properties of the cellulose from which it is made. Over the past 100 years or so the viscose process has undergone many refinements. However, the basic chemistry is still the same. Through this route short-fibre cellulose (wood pulp) is converted in a series of controlled and coordinated steps to a spinnable solution (dope) and then into longer filaments which may be precisely controlled in terms of length, denier, physical properties and cross-sectional shape.

The rayon filaments are formed when the viscose solution is extruded through the very small holes of a spinneret into a spin bath consisting basically of sulphuric acid, sodium sulphate, zinc sulphate and water. The spinbath often also contains a low level of surfactant. Coagulation of the filaments occurs immediately upon neutralising and acidifying the cellulose xanthate followed by controlled stretching and decomposition of the cellulose xanthate to cellulose. These latter steps are important for obtaining the desired tenacity and other properties of the fibre. Finally, the newly formed filaments are washed free of acid, chemically treated (desulphurised), and bleached prior to final washing and applying a processing finish. This is done either in the form of continuous filament (yarns or tow) or as cut staple, prior to drying and packaging.

生词与词组

1.pulp [pʌlp]v. 打浆；制浆
2.alkaline [ˈælkə,laɪn]adj. 碱性的；碱的
3.sodium hydroxide 氢氧化钠；烧碱
4.sodium sulphide 硫化钠；硫化碱
5.delignification [dɪlɪgnɪfɪˈkeɪʃ(ə)n]n. 去木质素

6.bleaching [ˈbliːtʃɪŋ]*n.* 漂白 *adj.* 漂白的

7.extractible [ɪkˈstræktəbl]*adj.* 可提取的；可抽出的

8.viscose [ˈvɪskəʊs]*n.* 黏胶；黏胶纤维；黏胶丝；黏胶液

9.dope [dəʊp]*n.* 黏稠物；胶状物

10.burgeoning [ˈbɜː(r)dʒ(ə)nɪŋ]*adj.* 迅速发展的；迅速成长的；生机勃勃的

11.rayon [ˈreɪɒn]*n.* 人造丝；人造纤维

12.refinement [rɪˈfaɪnmənt]*n.* 改进；改良

13.cross-sectional shape 横截面形状

14.spinneret [ˈspɪnəˌret]*n.* 喷丝头

15.spin bath 凝固浴；自旋浴

16.sulphuric acid 硫酸

17.zinc sulphate 硫酸锌

18.sodium sulphate 硫酸钠

19.surfactant [sɜː(r)ˈfæktənt]*n.* 表面活性剂；表面活性物质

20.neutralising [ˈnjuːtrəˌlaiz]*n.* 中和；中和作用 *adj.* 中和的

21.xanthate [ˈzænˌθeɪt]*n.* 磺酸盐；黄原酸盐

22.acidify [əˈsɪdɪfaɪ]*v.* 酸化

23.decomposition [ˌdiːkɒmpəˈzɪʃ(ə)n]*n.* 分解；降解

24.desulphurise [diːˈsʌlfjʊˌraɪz]*v.* 脱硫；使脱硫

25.tow [təʊ]*n.* 丝束

26.extrude [ɪkˈstruːd]*v.* 挤压；挤压成型

译文

第 13 课　工业纤维素

1. 制浆法

尽管一些国家使用了非木制的其他纤维材料，但化学和物理法制备的木质浆原材料在世界范围内的纸板行业中仍占主导地位。最常见的化学制浆工艺是硫酸盐法制浆工艺，这是一种以氢氧化钠和硫化钠作为活性脱木素剂的碱性工艺。该制浆工艺可处理包括软木和硬木在内的各种各样纤维原材料。

2. 溶解制浆工艺

制浆过程必须溶解大部分木质素并改性残留的木质素以成功实现漂白。漂白可去除残留部分木质素,并增加浆液的亮度和洁净度。最终结果是一种工艺纤维素尽可能地从木质素和半纤维素中分离、提取。通过制浆和漂白工序,最终溶解法制浆的 α- 纤维素含量通常为 90%-96%。

3. 黏胶工艺

1893 年,黏胶工艺的第一项专利授予了英国的克劳斯(Cross)和贝文(Bevan)。到 1908 年,黏胶纺丝已成为蓬勃发展的纺织工业的重要组成部分。在所有人造纤维中,黏胶(或人造丝)仍作为应用最广泛的人造纤维而占据独有地位。这是因为利用纤维素的特性,在制造时就可通过化学方法和结构性方法来设计纤维。这是因为能利用纤维素的特性,在制造中,通过化学和结构上对纤维进行改造。在过去 100 多年间,黏胶工艺经历了许多改进。然而,基本的化学反应仍未变化。通过这条路线,短纤维纤维素(木浆)在一系列控制和协同的操作下被转化为可纺溶液(原液),然后再转化为更长的长丝,这些长丝可在长度、旦数、物理特性和横截面形状上进行精确控制。

黏胶溶液由喷丝头的细孔中挤压出,在硫酸、硫酸钠、硫酸锌和水组成的纺丝凝固浴中形成人造长丝。纺丝凝固浴常含有一些低浓度的表面活性剂。通过控制伸长和磺酸酯纤维素分解纤维素速率,在纤维素磺酸酯的中和、酸化过程中获得凝固长丝。接下来的步骤对纤维获得理想的韧性和其他性能至关重要。最后,新生长丝经无酸洗涤、化学处理(脱硫)、漂白,再进行最终洗涤和加工整理。以连续长丝(纱线或丝束)或割断纤维加工后,再干燥,打包。

Lesson 14   Lyocell

Lyocell is the first in a new generation of cellulosic fibres. The development of lyocell was driven by the desire for a cellulosic fibre which exhibited an improved cost/performance profile compared to viscose rayon. The other main driving force was the continuing demands for industrial processes to become more environmentally responsible and utilise renewable resources as their raw materials. The resultant lyocell fibre meets both demands. Lyocell was originally conceived as a textile fibre. The first commercial samples were produced in 1984 and fibre production has been increasing rapidly ever since. Fabrics made from lyocell can be engineered to produce a wide range of drapes (how the fabric hangs), handles (how the fabric feels) and unique aesthetic effects. It is very versatile and can be fabricated into a wide range of different fabric weights from women's lightweight blouse fabric through to men's suiting.

Other end-uses, such as nonwoven fabrics and papers, are being developed. These non-textile end-uses will become progressively more important as the special properties of lyocell fibres enables products with enhanced performance characteristics to be developed. Lyocell is a 100% cellulosic fibre derived from wood-pulp produced from sustainable managed forests.The wood-pulp is dissolved in a solution of hot N-methyl morpholine oxide. The solution is then extruded (spun) into fibres and the solvent extracted as the fibres pass through a washing process. The manufacturing process is designed to recover over 99% of the solvent, helping minimise the effluent. The solvent itself is nontoxic and all the effluent produced is nonhazardous. The direct dissolution of the cellulose in an organic solvent without the formation of an intermediate compound differentiates the new generation of cellulosic fibres, including lyocell, from other cellulosic fibres such as viscose. This has led to the new generic name "lyocell" being accepted for labelling purposes.

Lyocell has all the benefits of being a cellulosic fibre, in that it is fully biodegradable, it is absorbent and the handle can be changed significantly by the use of enzymes or chemical finishing techniques. It has a relatively high strength in both the wet and dry states which allows for the production of finer yarns and lighter fabrics. The high strength also facilitates its use in various mechanical and chemical finishing treatments both under conventional and extreme conditions. The physical characteristics of lyocell also result in its excellent blending characteristics with fibres such as linen, cashmere, silk and wool.

生词与词组

1.aesthetic [i:s'θetɪk]*n.* 美感　*adj.* 美感的；美的

2.N-methyl morpholine oxide N-甲基吗啉

3.effluent ['efluənt]*n.* 污水；废水；排出物

4.nontoxic [nɒn'tɒksɪk]*adj.* 无毒的

5.nonhazardous [ˌnɒn'hæzədəs]*adj.* 无危害的；安全的

6.intermediate compound 中间化合物

7.enzyme ['enzaɪm]*n.* 酶；酶法；酶制剂

译文

第 14 课　莱赛尔纤维

莱赛尔纤维是第一代纤维素纤维。与黏胶人造纤维相比，莱赛尔纤维的发展得益于纤维素纤维在成本和性能方面的改进。对工业生产更环保和使用可再生资源的原材料的持续性需求也是莱赛尔纤维发展的另一主动力。所制的莱赛尔纤维满足这两个要求。莱赛尔纤维最初可作为一种纺织纤维。1984 年第一个商业化样品出现,此后纤维产量一直在快速增长。以莱赛尔设计生产的织物可获得多样的悬垂性(织物悬垂如何)、手感(织物感觉如何)以及独特的美感。它用途广泛,可以制成各种不同重量的面料,从女式轻质上衣面料到男式西装。

其终端使用正在开发中,如非织造布和纸。因莱赛尔纤维的特殊性质赋予了其更佳的产品性能,这在非纺织品方面会变得日益重要。莱赛尔纤维是一种源于再生林木浆的纯纤维素纤维(100%)。木浆溶解在热的 N-甲基吗啉氧化物溶液中。溶解液再被挤出成(纺成)纤维,在纤维洗涤时萃取溶剂。设计的生产工艺使得超过 99% 的溶剂被回收,这有助于最大限度地减少废水排放。溶剂本身是无毒的,所有废水是无害的。新一代纤维素纤维(包括莱赛尔)与其他纤维素纤维(如黏胶)区别在于纤维素在有机溶剂中直接溶解而不形成中间化合物。这使得以新通用名"莱赛尔"作为标签被接受。

莱赛尔纤维具有纤维素纤维所有的优点,可完全生物降解,吸水性强,以及经酶或化学整理技术可大幅改良手感,在湿态和干态下它具有相对较高的强度,也使得其可生产更细的纱线和更轻的织物。高强度利于它在常规和极端条件下进行各种机械和化学整理。莱赛尔的物理特性也使其可与亚麻、羊绒、丝绸和羊毛等纤维进行混纺。

Lesson 15   Polyester fibres

Polyester fibres, and by this we mean largely poly ethylene terephthalate (PET) fibres(see Fig. 15.1), dominate the world synthetic fibres industry. They constitute, by a considerable margin, the largest volume of synthetics and far outweigh nylons, rayon and acrylic fibres. They are inexpensive, easily produced from petrochemical sources, and have a desirable range of physical properties.

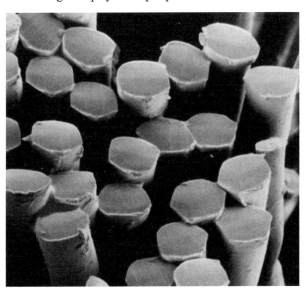

Figure 15.1   PET fibres

They are strong, lightweight, easily dyeable and wrinkle-resistant, and have very good wash-wear properties. Both as continuous filament yarn and staple fibre, they are used in countless varieties, blends and forms of textile apparel fibres, household and furnishing fabrics. They form microfibres for outdoor wear and sportswear. Polyesters are used in carpets, industrial fibres and yarns for tyre cords, car seat belts, filter cloths, tentage fabrics, sailcloth and so on. While the dominant polymer is PET, other polyesters also have their place. Polybutylene terephthalate (PBT) and lately polytrimethylene terephthalate (PTT) are used in polyester carpet fibre because of their superior fibre resilience. Biodegradable polyesters derived from lactic acid, glycolic acid and other aliphatic hydroxyacids are used in medical appliances (e.g., dissolvable sutures and drug delivery polymer vehicles). The latest innovations are cheap, biodegradable, polylactide fibres made from lactic acid produced by fermentation of cornstarch. This is an essentially "green" chemical process, and the materials are aimed at disposable products that will quickly biodegrade and constitute no threat to the environment.

1.Classification

Polyesters are broadly classified into two types for thermoplastic cross-linked thermosets and elastomers. Thermoplastic polyesters can be further classified into film-forming and fibre-forming polyesters. Linear aromatic commercial polyesters are classified into two types based on the type of aromatic moiety present in the polyester main chain. They are (i) phthalates and (ii) naphthalates. Polyethylene terephthalate (PET), polytrimethylene terephthalate (PTT), polybutylene terephthalate (PBT) belong to the phthalate group because these polyesters were derived from purified terephthalic acid (PTA) or dimethyl terephthalate (DMT).

2.Poly butylene terephthalate (PBT)

A more accurate name for this polymer would be poly-tetramethylene terephthalate, but "PBT" is a well-established acronym in the engineering moulding resins business. Originally developed as a textile fibre, it later became much more important as a highly crystallisable injection moulding polymer and an engineering resin. Nevertheless it maintains a fibre presence in the market, chiefly as bulked continuous filament (BCF) carpet fibre, where its resilience is important.

3.Poly trimethylene terephthalate (PTT or PPT)

Whinfield and Dickson in 1941 discovered poly trimethylene terephthalate along with PET and PBT. The material was recognised for many years by textile chemists as a fibre-forming polymer affording fibres that had excellent physical properties and outstanding resilience. The lack of an economic source of pure 1,3-propanediol (PDO) was an insuperable obstacle for many years, but recently the situation has changed entirely and 3GT fibres are now a commercial fact. There seems to be a little confusion on an acronym for this polyester. Shell prefers to use the term PTT (poly trimethylene terephthalate) but DuPonts call it PPT (poly propylene terephthalate).

4.Modification of polyester fibres

Plain PET fibre on a yarn package is unlikely to suit everyone of the very many different markets for which it is used. It is often necessary to modify the fibre in various ways. This is a broad topic, covering both chemical and physical modifications to both polymer and fibre.

5.Dyeing polyesters

When PET first appeared on the market, it caused many new problems for traditional dyers, since it had no functional groups to give it any affinity for the usual dyestuffs. Natural fibres such as wool, cotton and silk (and later nylon) were well understood and they had

good dye affinities, owing to multiple fibre functionalities such as
$-NH_2$, -COOH and -OH. The only way to dye polyester was to first
force a dye into the fibre and then rely on van der Waals forces to
hold the dye in place. Classic cationic and anionic dyes for wool and
silk or direct dyes for cotton all had water-solubilising groups such
as $-NR_3^+$ and $-SO_3-$ groups. Such dyes had little or no affinity for the
hydrophobic PET.

生词与词组

1.polyester [ˌpɒliˈestə(r)]n. 聚酯（纤维）；涤纶

2.poly ethylene terephthalate 聚对苯二甲酸乙二醇酯

3.acrylic [əˈkrɪlɪk]n. 丙烯酸纤维；腈纶 adj. 丙烯酸的

4.nylon [ˈnaɪlɒn]n. 尼龙

5.wash-wear [ˌwɒʃˈweə(r)]adj. 快干的；洗可穿的

6.furnishing [ˈfɜ:(r)nɪʃɪŋ]n. 家饰

7.microfibre [ˈmaɪkrəʊˌfaɪbə(r)]n. 微纤维；超细纤维

8.sportswear [ˈspɔ:(r)tsˌweə(r)]n. 运动服装

9.filter cloth 滤布

10.tentage fabric 帐篷织物

11.poly butylene terephthalate 聚对苯二甲酸丁二醇酯

12.lactic acid 乳酸

13.glycolic acid 羟基乙酸

14.aliphatic hydroxyacid 脂肪族羟基酸

15.polylactide [ˌpɒliˈlæktaid]n. 聚乳酸

16.fermentation [ˌfɜ:menˈteɪʃ(ə)n]n. 发酵；发酵法

17.cornstarch [ˈkɔ:(r)nˌstɑ:(r)tʃ]n. 玉米淀粉

18.thermoplastic [ˌθɜ:məʊˈplæstɪk]adj. 热塑性的；热塑的

19.thermoset [ˈθɜ:məʊˌset]n. 热固性 adj. 热固（硬）性的

20.film-forming 成膜

21.fibre-forming 纤维成形

22.elastomer [ɪˈlæstəmə]n. 弹性体；高弹体

23.aromatic [ˌærəˈmætɪk]n. 芳香族；芳香 adj. 芳香族的；芳香的

24.phthalate [ˈθæleit]*n.* 邻苯二甲酸

25.dimethyl terephthalate 聚对苯二甲酸二甲酯

26.acronym [ˈækrənɪm]*n.* 首字母缩略词

27.moulding [ˈməʊldɪŋ]*n.* 造型；模制

28.bulked continuous filament 膨体连续长丝

29.1,3-propanediol 1,3-丙二醇

30.insuperable obstacle 不可克服的障碍

31.dyestuff [ˈdaɪstʌf]*n.* 染料；着色剂

32.affinity [əˈfɪnəti]*n.* 亲和力；亲和度

33.van der Waals 范德华力

34.cationic [ˌkætɪˈəʊnɪk]*adj.* 阳离子的；阳离子型的；阳离子化的

35.legendary [ˈledʒəndri]*adj.* 非常著名的；传奇的

36.sailcloth [ˈseɪlklɒθ]*n.* 帆布

译文

第 15 课　聚酯纤维

聚酯纤维,我们在这里主要指的是聚对苯二甲酸乙二醇酯（PET）纤维(见图 15.1),其主导着世界合成纤维行业。它是合成纤维产量中占比最大的,远远超过尼龙、人造丝和丙烯酸纤维。聚酯纤维价格低廉,易从石油资源中获得,具有一系列理想的物理性质。

它们强度高,质量轻,易染色,抗皱,且具有良好的洗可穿性。以连续长丝纱和短纤维,它们被制成无数品种、混纺和形式的纺织服装纤维、家居和装饰织物。它们制成超细纤维用于户外服装和家具织物。

聚酯纤维也用于地毯、轮胎帘子布、汽车座椅安全带、过滤布、帐篷织物、帆布等的工业纤维和纱线。虽然主要聚合物是 PET,但其他聚酯也有一席之地。聚对苯二甲酸丁二醇酯（PBT）和最新的聚对苯二甲酸丙二醇酯（PTT）因其优良的纤维回弹性而被用于聚酯类地毯纤维。由乳酸、乙醇酸和其他脂肪族羟基酸合成的生物可降解聚酯,可用于医疗器械(如可溶解的缝合线和药物传输聚合物载体)。由玉米淀粉发酵的乳酸可合成新型廉价、可生物降解的聚乳酸纤维。这是一种"绿色"化

学工艺,其旨在生产可快速生物降解,并且对环境不构成威胁的一次性材料。

图 15.1　PET 纤维

1. 分类

　　聚酯大致分为热塑性交联热固性体和弹性体两类。热塑性聚酯可进一步分为成膜聚酯和成纤聚酯。基于聚酯主链中芳香族基团类型,商用线型芳香族聚酯分为邻苯二甲酸酯和萘二甲酸酯两类。聚对苯二甲酸乙二醇酯(PET)、聚对苯二甲酸丙二醇酯(PTT)、聚对苯二甲酸丁二醇酯(PBT)属于邻苯二甲酸酯组,因为这些聚酯由纯化的对苯二甲酸(PTA)或聚对苯二甲酸二甲酯(DMT)衍生而来。

2.PBT 纤维

　　这种聚合物更准确的名称是聚对苯二甲酸丁二醇酯,但"PBT"是工程成型树脂的首字母缩写词。最初作为纺织纤维,后来它更多作为高结晶度的注塑聚合物和工程树脂使用。尽管它仍在市场上占有一席之地,但主要作为地毯用的高弹性膨体长丝(BCF)。

3.PTT 或 PPT 纤维

菲尔德（Whinfield）和迪克生（Dickson）在 1941 年发现了聚对苯二甲酸丙二醇酯以及 PET 和 PBT。多年来，该材料被纺织化学家认为是一种成纤聚合物，其具有优异物理性能和出色回弹性。多年来，纯 1,3-丙二醇（PDO）产品匮乏，一直是一个无法克服的障碍，但近期这已经完全转变，3GT 纤维已可商业化生产。这种聚酯的首字母缩写词似乎有些混乱。壳牌公司更喜欢用 PTT（聚对苯二甲酸三甲酯）这个术语，而杜邦公司则称之为 PPT（聚对苯二甲酸丙二醇酯）。

4. 改性聚酯纤维

筒装的普通 PET 纤维不适于所有非常不同的市场使用，需要经常以各种方法对纤维进行改性。这是一个广泛的主题，涵盖了聚合物及纤维的化学和物理改性。

5. 聚酯染色

当聚酯首次出现在市场上时，因其官能团与常规染料无亲和力，给传统染料带来了许多新问题。天然纤维如羊毛、棉和丝(以及后来的尼龙），因具有 $-NH_2$、$-COOH$ 和 $-OH$ 等官能团，使它们具有良好的染料亲和性。聚酯染色的唯一方法是先将染料渗入纤维中，再依靠范德华力将染料固定在合适的位置上。羊毛和丝的阳离子、阴离子染料，或棉的直接染料都含有水溶性基团，如 $-NR_3^+$ 和 $-SO_3^-$，而这些染料与疏水性涤纶的亲和性很小甚至没有。

Lesson 16 Nylon fibres

Nylon was the first synthetic fibre to go into full-scale production and the only one to do so prior to World War II. The development arose from the work of Wallace Carothers for DuPont starting in 1928. By 1935 the first nylon 6,6 polymers had been prepared and pilot plant production started in 1938. In the following year, the first plant for nylon fibres went into production and the first stockings went on sale in October 1939. The first production in the UK was under licence from DuPont by British Nylon Spinners, a company jointly formed by Courtaulds and ICI, and started in 1941. Most production during World War II was devoted to military uses, particularly for parachute fabrics. Only in 1946 did the fibre start to be available for domestic uses.

In 1950 the total world production of synthetic fibres was only 69,000 tonnes, and almost all of this was nylon. In 1970 nylon accounted for 40% of the total synthetic fibre production with just under 2×106 tonnes. The applications also expanded from the initial hosiery market to reinforcement of rubber in tyres and belts, and to carpets, often in blends with wool. The easy-care properties of the fibre were exploited in its use in underwear, bedsheets, shirts and other apparel. However, in many of these textile applications, flat yarns with relatively coarse filaments were used in tightly woven or knitted fabrics.

These fabrics had a limp and rather plastic handle, poor ability to wick moisture away from the body, and a strong tendency to build up static charges, leading to clinging and sparking. In the fashion industry the word "nylon" became almost pejorative.

The world production of nylon has continued to increase slowly and now exceeds 4×106 tonnes per annum. Producers have marketed the fibre for applications in which its properties are best utilised. These include carpets, tights and stockings, tyres and belts. For apparel, producers engineer their products to meet the needs of particular uses — filaments have become finer and frequently non-circular in cross section(see Fig. 16.1); textured yarns are often used. These products, together with changes in fashion, led in the late 1990s to a revival of interest in nylon for outerwear.

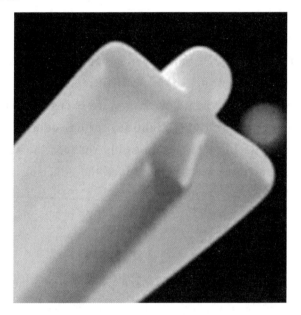

Figure 16.1　Non-circular cross section of fiber

1.Fibre modification

Indeed the term "fibre modification" implies that some fibres have no modifications at all, being extruded from pure nylon 6 or 6,6 from a circular spinneret with no additives, and not subjected to any form of after treatment.

The advent of nylon fibres enabled the production of continuous filaments which, when twisted to make yarns, gave sheer, shiny and smooth fabrics. However, the universal shininess of such fabrics led to experiments with delustrants, which were used initially with viscose continuous filament yarns in the 1930s. The need for an effective delustrant became apparent with the production of the first nylons for general apparel applications in the early 1950s. Fabrics have a lustre that is only attractive in particular applications and tend to be excessively translucent. The fabrics are even more transparent when wet. The first nylon swimsuits had considerable advantages over the previously used woollen costumes, which could weigh as much as 5kg when wet. They were strong, light, well fitting and quick drying, but the difficulty was that when wet, they were excessively transparent. The presence of small, highly light-scattering, delustrant particles of titanium dioxide in the fibres not only reduces the surface lustre but increases the fibre opacity and hence the covering power of the derived textile.

2.Bicomponent fibres

Bicomponent technology has been used to introduce functional and novelty effects, including stretch, to nylon. Bicomponent fibres have two distinct polymer components, usually of the same generic class in the fibre cross-section, e.g., one part of the fibre could be nylon 6,6 and the other nylon 6. They are also described as "conjugate fibres", particularly by some Asian producers. If the components are from different generic classes, the fibre is sometimes said to be biconstituent. In bicomponent and biconstituent fibres, each component is normally fibreforming. The spinneret and the polymer feed channels can be designed so that the two polymer streams meet just prior to the spinneret hole and emerge side by side or as a sheath and core. The fibres then need to be drawn and processed in the usual way. The engineering required to ensure that each hole of a spinneret receives a feed of two polymers at the required rate is complex, but the difficulties have been overcome and there are a large number of bicomponent nylon fibres in commercial production. The ratio of the two polymeric components need not be 50/50 and can be varied according to the application.

生词与词组

1.military [ˈmɪlətri]*adj.* 军事的；军用的

2.parachute [ˈpærəʃuːt]*n.* 降落伞

3.hosiery [ˈhəʊziəri]*n.* 袜子；袜业

4.reinforcement of rubber 改性橡胶

5.tyre [ˈtaɪə]*n.* 轮胎

6.belt [belt]*n.* 皮带；腰带

7.easy-care [ˈiːzi keə(r)]*adj.* 免熨烫的

8.bedsheet [ˈbedˌʃiːt]*n.* 床单

9.limp [lɪmp]*adj.* 柔软的；易弯曲的

10.wick [wɪk]*v.* 毛细管作用

11.clinging [ˈklɪŋɪŋ]*adj.* 紧身的；贴身的

12.pejorative [pɪˈdʒɒrətɪv]*adj.* 贬义的　*n.* 贬义词

13.stocking [ˈstɒkɪŋ]*n.* 长筒袜；丝袜

14.non-circular [ˌnəʊn ˈsɜːkjələ(r)]*adj.* 非圆形的；非循环的；非周期的

15.textured yarn 变形纱线；变形纱

16.delustrant [diːˈlʌstrənt]*n.* 消光剂；除光剂

17.sheer [ʃɪə(r)]*adj.* 极薄的；透明的

18.translucent [trænzˈluːsnt]*adj.* 半透明的

19.swimsuit [ˈswɪmsuːt]*n.* 泳衣

20.light-scattering [laɪt ˈskætərɪŋ]*n.* 光散射

21.titanium dioxide 二氧化钛

22.opacity [əʊˈpæsəti]*n.* 不透明；模糊

23.bicomponent [ˌbaikəmˈpəunənt]*n.* 双组分

24.conjugate [ˈkɒndʒəgeɪt]*adj.* 结合的

25.biconstituent [ˌbaɪkənˈstɪtjʊənt]*adj.* 双成分的

26.sheath [ʃi:θ]*n.* 鞘；壳；外皮

27.silk-like [ˈsɪlk laɪk]*adj.* 仿丝的；像丝绸一样的

译文

第16课　尼龙纤维

尼龙是第一种大规模生产的合成纤维，也是第二次世界大战前唯一一种规模生产的合成纤维。这一发明源于1928年杜邦公司的华莱士·卡罗瑟斯（Wallace Carothers）。1935年，第一批尼龙6,6聚合物试制完成；1938年，中试装置开始生产。次年，第一家尼龙纤维工厂投产，第一批长筒袜于1939年10月上市销售。1941年，由Courtaulds和ICI联合成立的英国尼龙纺纱公司（British Nylon Spinners）获得杜邦公司（DuPont）的授权，开始在英国生产尼龙。在第二次世界大战期间，其进行大批量生产，用于军事用途，特别是降落伞面料。直到1946年，这种纤维才开始家用。

1950年世界合成纤维的总产量只有69000吨，几乎全部是尼龙纤维。1970年，尼龙占合成纤维总产量的40%，仅略低于2×106吨。其应用也从最初的袜子市场扩展到轮胎、皮带的橡胶增强材料，以及与羊毛混纺的地毯。该纤维易护理，也被用于内衣、床单、衬衫和其他服装。然而，在许多纺织应用中，相对较粗扁丝用于制造较紧的机织或针织织物。

这些织物具有类似塑料的柔软手感，较差的吸湿能力，并且其极易集聚电荷而贴体，生电。在时尚领域，"尼龙"这个词几乎带有贬义。

世界尼龙产量持续缓慢增长，现已超过4×106吨／年。生产商已将这种纤维用于最能充分利用其特性的场合，这包括地毯、紧身衣和长袜、轮胎和皮带。在服装领域，生产商设计了满足特殊用途的产品，如更细的长丝、非圆形横截面（见图16.1）、变形纱。这些产品合力转变了时尚，在20世纪90年代末出现了尼龙外套的兴起。

图 16.1　非圆形截面纤维

1. 纤维改性

事实上,"纤维改性"术语意味着一些纤维几乎未改性,由圆形喷丝头挤出的纯尼龙 6 或尼龙 6,6 不含添加剂,不进行任何形式的后处理。

新问世的尼龙纤维可加工成长丝,再加捻成纱线,可制得轻薄、有光泽和光滑的织物。然而,这种织物的大量光泽推动了消光剂的试用,并在 1930 年首次应用于黏胶长丝纱。1950 年初,第一批服用尼龙的生产,对有效消光剂的需求显著增加。织物的光泽仅在特定应用中才具有吸引力,往往过于透明。这种面料在湿态更加透明。与湿重达 5 千克的羊毛泳衣相比,第一代尼龙泳衣具有相当大的优势。它们高强,质轻,合身且快干,但难点在于其遇湿过于透明。将细微、高光散射、消光的二氧化钛颗粒添加至纤维中,其不仅可降低表面的光泽,还可增强纤维的不透明度,从而提升织物的覆盖能力。

2. 双组分纤维

双组分技术向尼龙中引入了功能性和新颖性效果(包括伸长性)。双组分纤维是纤维横截面内有两种不同的聚合物成分,如纤维的一部分为尼龙 6,6,另一部分为尼龙 6。一些亚洲生产商称它们为"复合纤维"。

如果成分来自不同的类别,有时也称为"双成分纤维"。在双组分和双成分纤维中,每个组分都是可成纤的。两股聚合物流在喷丝头前端相遇,以并排或皮芯的形式通过设计的喷丝头和聚合物进料通道。然后,纤维需要以常规方式进行拉伸和处理。确保喷丝头的每个孔以预定速率接收两种聚合物的工程很复杂,但难题已克服,大量的双组分尼龙纤维已商业生产。两种聚合物组分的比例不必为 50/50,其可依据应用而变化。

Lesson 17   Acrylic fibres

The definition of an acrylic fibre is one that contains at least 85% by mass of acrylonitrile comonomer in the polymer chain. All commercial processes for the manufacture of acrylic fibres are based on free radical polymerisation, as this technique gives the required combination of polymerisation rate, ease of control and properties such as whiteness, molecular weight and linearity. This polymerisation technique also allows the incorporation of other comonomers, which impart important fibre processing properties, and in most cases, dye sites for colouration purposes. Acrylic fibres vary from virtual homopolymers of acrylonitrile, which tend to be used for industrial or high-performance related end-uses, to fibres containing up to 15% of comonomers for more typical textile end-uses.

生词与词组

1.flat yarn 扁纱
2.acrylonitrile [ˌækrɪləʊˈnaɪtraɪl]*n.* 丙烯腈；丙烯
3.homopolymer [ˌhɒməʊˈpɒlɪmə]*n.* 均聚物；同聚物

译文

第 17 课　腈纶

丙烯酸纤维的定义是在聚合物链中含有至少 85%（质量）丙烯腈共聚单体。所有丙烯酸纤维的制造工艺都基于自由基聚合，因该技术提供了所需的聚合速率、易于控制和特性（如白度、分子量和线性度）的组合。这种聚合技术还允许加入其他共聚单体，其赋予了重要的纤维加工性能，在大多数情况下为染色位点。从用于产业或高性能场合的全均聚丙烯腈体纤维到含有 15% 共聚单体的常规纺织用途纤维均为丙烯酸纤维。

Lesson 18　High-performance fibres

In a sense, all fibres except the cheapest commodity fibres are high-performance fibres. The natural fibres (cotton, wool, silk...) have a high aesthetic appeal in fashion fabrics (clothing, upholstery, carpets...). Until 100 years ago, they were also the fibres used in engineering applications—what are called technical or industrial textiles. With the introduction of manufactured fibres (rayon, acetate, nylon, polyester...) in the first half of the twentieth century, not only were new high-performance qualities available for fashion fabrics, but they also offered superior technical properties. For example, the reinforcement in automobile tyres moved from cotton cords in 1900, to a sequence of improved rayons from 1935 to 1955, and then to nylon, polyester and steel, which dominate the market now. A similar replacement of natural and regenerated fibres by synthetic fibres occurred in most technical textiles.

1.Aramids

Aromatic polyamides became breakthrough materials in commercial applications as early as the 1960s, with the market launch of the meta-aramid fibre Nomex® (Nomex® is a DuPont registered trademark), which opened up new horizons in the field of thermal and electrical insulation. A much higher tenacity and modulus fibre was developed and commercialised, also by DuPont, under the trade name Kevlar® (Kevlar® is a DuPont registered trademark) in 1971. Their outstanding potential derived mostly from the anisotropy of their superimposed substructures presenting pleated, crystalline, fibrillar and skin-core characteristics. Aramid fibres have unique properties that set them apart from other fibres. The tensile strength and modulus of aramid fibres are significantly higher than those of earlier organic fibres, and fibre elongation is lower. Aramid fibres can be woven on fabric looms more easily than brittle fibres such as glass, carbon or ceramic.

2.Gel-spun high-performance polyethylene fibres

Gel-spun polyethylene fibres are ultra-strong, high-modulus fibres that are based on the simple and flexible polyethylene molecule. They are called high-performance polyethylene (HPPE) fibres, high-modulus polyethylene (HMPE) fibres or sometimes extended chain polyethylene (ECPE) fibres. The gel-spinning process uses physical processes to make available the high potential mechanical properties of the molecule. Owing to low density and good mechanical properties, the performance on a weight basis is extremely high. The chemical nature of polyethylene remains in the gel-spun fibre and this can both be positive and a limitation: abrasion, flexlife, etc. are very high but the melting point is sometimes too low for certain applications. The titre of the monofilaments varies from about 0.3 denier per filament (dpf) (0.44dtex) to almost 10dpf (11dtex). Tenacity of one filament may well be over 5N/tex, and the modulus can be over 120N/tex. Staple fibre is not produced as such. Stretch broken and cut fibres are used by specialised companies. Most fibre grades have a more or less circular cross-section.The fibre skin is smooth. The fibre is highly crystalline; the crystallinity is typically over 80%. The larger part of the non-crystalline fraction is in the form of an interphase that is characterised by a high density, a high orientation and restricted mobility of the molecular chains.

3.Carbon fibres

Carbon fibres have been under continuous development for the last 50 years. There has been a progression of feedstocks, starting with rayon, proceeding to polyacrylonitrile (PAN), on to isotropic and mesophase pitches, to hydrocarbon gases, to ablated graphite and finally back to carbon-containing gases. PAN-based fibre technologies are well developed and currently account for most commercial production of carbon fibres. "General purpose" fibres made from isotropic pitch have modest levels of strength and modulus. However, they are the least expensive pitch-based fibre, and are useful in enhancing modulus or conductivity in many applications. PAN-based fibres are the strongest available; however, when they are heat treated to increase modulus, the strength decreases. Mesophase pitch fibres may be heat treated to very high modulus values, approaching the in-plane modulus of graphite at 1 TPa.

4.Glass fibres

The drawing of glass into fine filaments is an ancient technology, older than the technology of glass blowing. Winding coarse glass fibres onto a clay mandrel was used as an early manufacturing route for a vessel. With the advent of glass blowing, similar fibre technologies were used to decorate goblets.

In the 1700s, Réaumur recognised that glass could be finely spun into fibre that was sufficiently pliable to be woven into textiles. Napoleon's funeral coffin was decorated with glass fibre textiles. By the 1800s, luxury brocades were manufactured by co-weaving glass with silk, and at the Columbia Exhibition of 1893, Edward Libbey of Toledo exhibited dresses, ties and lamp-shades woven from glass fibre.

The scientific basis for the development of the modern reinforcing glass fibre stems from the work of Griffiths, who used fibre formation to validate his theories on the strength of solids. Glass fibres are used in a number of applications which can be divided into four basic categories: insulations, filtration media, reinforcements, and optical fibres.

生词与词组

1.gel-spun 凝胶纺纱

2.upholstery [ʌpˈhəʊlstəri]*n.* 内饰；面料；蒙皮材料

3.meta-aramid [ˌmetəˈærəmɪd]*n.* 间位芳纶

4.superimposed substructure 叠加子结构

5.fibrillar [ˈfaɪbrɪlə]*adj.* 纤丝状的；纤维状的

6.skin-core [ˈskɪnˌkɔː(r)]*n.* 皮芯；皮芯层

7.loom [lu:m]*n.* 织机；织布机

8.ceramic [səˈræmɪk]*n.* 陶瓷 *adj.* 陶瓷的

9.gel-spun polyethylene 凝胶纺丝聚乙烯

10.ultra-strong [ˈʌltrə-strɒŋ]*adj.* 超强的

11.flexlife [ˈfleksləɪf]*n.* 屈折年限；挠曲寿命；挠性寿命

12.melting [ˈmeltɪŋ]*n.* 熔化；软化

13.multifilament [ˌmʌltɪˈfɪləmənt]*n.* 复丝；多纤维 *adj.* 多股的；多纤维的

14.titre [ˈtaɪtə]*n.* 纤度

15.monofilament [ˌmɒnəˈfɪləmənt]*n.* 单丝；单纤

16.feedstock [ˈfiːdˌstɒk]*n.* 原料

17.isotropic [ˌaɪsəʊˈtrɒpɪk]*adj.* 各向同性的

18.mesophase pitch 中间相沥青

19.hydrocarbon gas 碳氢化合物气体；烃类气体

20.ablated graphite 烧蚀石墨

21.carbon-containing gas 含碳气体

22.pitch-based fibre 沥青基纤维

23.conductivity [ˌkɒndʌkˈtɪvəti]*n.* 传导率；导电率

24.mesophase [ˈmesəʊfeɪz]*n.* 中间相；中介相；介稳相

25.glass blowing 玻璃吹制

26.clay mandrel 黏土芯轴

27.vessel [ˈvesl]*n.* 器皿；容器

28.goblet [ˈɡɒblɪt]*n*. 高脚杯

29.coffin [ˈkɒfɪn]*n*. 棺材；棺椁；棺木

30.luxury brocade 豪华锦缎

31.lamp-shade 灯罩

32.filtration media 过滤介质；过滤材料

33.reinforcement [ˌriːɪnˈfɔːsmənt]*n*. 加强；强化；增强相

34.optical fibre 光纤

译文

第18课　高性能纤维

在某种意义上,除了最廉价的商品纤维外,所有纤维均属于高性能纤维。天然纤维(棉、羊毛、蚕丝等)用于时装面料(服装、室内装饰、地毯等),具有很高的艺术感染力。直到100年前,它们也用于工程应用——被称为技术或产业用纺织品。随着20世纪上半叶人造纤维(人造丝、醋酸纤维、尼龙、聚酯等)的诞生,其不仅赋予了时装面料新的性能,也提供了优越的技术性能。例如,汽车轮胎的增强材料从1900年的棉帘子线,到1935年至1955年的一系列改进人造丝,再到现在主导市场的尼龙、聚酯和钢铁。在多数产业用纺织品中,类似的合成纤维取代天然纤维和再生纤维时有发生。

1. 芳纶

1960年,随着间位芳纶纤维Nomex®(Nomex®是杜邦公司的注册商标)的上市,芳香族聚酰胺已成为商业应用的突破性材料,其在热、电绝缘方面开辟了新的领域。1971年,杜邦公司开发并商业化了一种高强高模的纤维,其商品名称为Kevlar®(Kevlar®是杜邦公司的注册商标)。它们的突出潜力源于其内部各向异性的子结构,呈现出堆叠、结晶、纤维状和皮芯的特征。芳纶纤维具有与其他纤维不同的独特性能。芳纶纤维的拉伸强度和模量明显高于早期有机纤维,且伸长率较低。与玻璃、碳或陶瓷等脆性纤维相比,芳纶纤维更容易在织机上机织造。

2. 高性能聚乙烯纤维

凝胶纺丝聚乙烯纤维由简单而灵活的聚乙烯分子合成,是一种超强高模量纤维。它们被称为高性能聚乙烯(HPPE)纤维、高模量聚乙烯(HMPE)纤维或延伸链聚乙烯(ECPE)纤维。凝胶纺丝工艺通过物理方法来获得分子潜在的高力学性能。由于聚乙烯密度低和机械性能良好,按重量计算其性能非常高。聚乙烯的化学性质在凝胶纺丝纤维中得以保持,这有优点也有缺点:磨损、弯曲寿命等非常高,但在一些场合其熔点太低。单丝的细度从 0.3 旦尼尔 / 丝(dpf)(0.44 分特)到 10dpf(11 分特)不等。单丝强力可达 5 牛厘 / 特以上,模量可达 120 牛厘 / 分特以上。而短纤维就不是这样生产的。长丝拉断或切断的纤维供特定公司使用。大多数纤维的截面或多或少呈圆形。纤维表面是光滑的。该纤维结晶度较高,通常超过 80%。以界面相形式存在的非结晶部分,具有高密度、高取向、不易移动的分子链。

3. 碳纤维

碳纤维已持续发展 50 年了。从人造丝开始,到聚丙烯腈(PAN),到各向同性和中间相沥青,到碳氢化合物气体,再到烧蚀石墨,最后再到含碳气体,碳纤维原料经历了一系列的发展历程。聚丙烯腈基碳纤维技术发展良好,是当前主流的碳纤维商业生产技术。由各向同性的石油沥青制成的"常规"碳纤维具有适中的强度和模量。然而,在诸多场合中,应用最廉价的沥青基纤维可提升产品的模量或导电性。聚丙烯腈基碳纤维强度极高,它们经热处理后模量上升而强度下降。中间相沥青基纤维经热处理后可获得极高的模量,接近 10^{12} 帕的石墨面内模量。

4. 玻璃纤维

玻璃拉丝是一项古老的技术,比吹制玻璃的技术还要古老。将粗玻璃纤维绕在黏土芯轴上是早期容器的制造方法。随着玻璃吹制技术的出现,类似的纤维技术也被用于装饰酒杯。

18 世纪,列奥溑尔(Réaumur)发现玻璃可以纺成更细的纤维,其具有足够的柔性,可织成织物。拿破仑的棺椁就是用玻璃纤维织物装饰的。到 19 世纪,用玻璃长丝和丝绸混合织成了精美的织锦;在 1893 年的哥伦比亚展览会上,来自托莱多(Toledo)的爱德华·利比(Edward

Libbey）展出了用玻璃纤维织成的裙子、领带和灯罩。

现代增强型玻璃纤维的科学基础源于格里菲思（Griffiths）的理论，他用纤维的成形物来验证相关的固体强度理论。玻璃纤维有多种用途，其可分为绝缘材料、过滤介质、增强材料和光学纤维四大类。

Lesson 19   Introduction to spinning technology

In 2005, approximately 62% of all textile fibres were processed into spun yarns (short and long staple), 8% into non-wovens and 30% into filament yarns, short-staple spinning (up to about 50mm) accounting for over 80% of all staple yarns spun, with cotton accounting for almost 70% of this. This is often referred to as the yarn manufacturing stage of the textile pipeline and as the "cotton or short-staple system". Yarn manufacturing in essence involves the following objectives:

- Sliver attenuation (drafting);
- Sliver evening (doubling, autolevelling);
- Fibre aligning and straightening;
- Fibre blending;
- Short fibre removal;
- Removal of foreign particles (also dust) and neps;
- Twist insertion;
- Winding, clearing (fault removal), waxing/lubrication (knitting yarns).

Ultimately the purpose of the preparatory processes and spinning is to convert into a yarn, at cost as effective as possible and with a minimum of waste, a relatively coarse cotton sliver, in which the fibres are individualised but fairly randomly arranged and also not always all that well blended and which contains undesirable short fibres, fibre hooks and foreign particles.

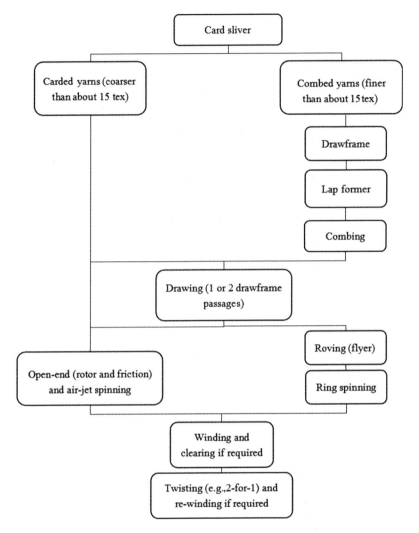

Figure 19.1   Typical cotton flow chart

The intention is to produce a yarn in which the fibres are as straight, orderly arranged and well aligned (parallel) as possible and which is as even as possible, both in appearance and composition, with the minimum number of imperfections, faults, trash, protruding hairs and short fibres. The yarn should also be on a package which is suitable for the subsequent fabric manufacturing processes.

In essence, two routes may be followed, the one for producing carded yarn and the other for producing combed yarn, the latter involving the additional process of combing (see Fig. 19.1).

生词与词组

1.sliver attenuation 条子拉细
2.imperfection [ˌɪmpəˈfekʃ(ə)n]*n.* 缺陷；不完全
3.drafting [ˈdrɑ:ftɪŋ]*n.* 牵伸
4.evening [ˈi:vnɪŋ]*adj.* 均匀的；一致的
5.doubling [ˈdʌblɪŋ]*n.* 加倍；成双
6.autolevelling [ɔ:təʊˈlevəlɪŋ]*n.* 自调匀整
7.fibre aligning and straightening 纤维对齐和校直
8.fibre blending 纤维混纺
9.twist insertion *n.* 加捻
10.hook [hʊk]*n.* 弯钩
11.trash [træʃ]*n.* 杂质；碎屑
12.protruding [prəˈtru:dɪŋ]*adj.* 突出的；伸出的

译文

第 19 课　纺纱技术简介

　　2005 年,大约 62% 的纺织纤维被加工纺成纱线(短纤和长纤),8% 被加工成无纺布,30% 被加工成长丝；短纤纺纱(长约 50 毫米)在所有纺纱中占比超过 80%,而其中棉纺纱占比近 70%。这通常被称为纺织流程的纱线制造阶段,也被称为"棉型或短纤维系统"。纱线生产本质上涉及以下目的：
- 条子拉细(牵伸)；
- 条子均匀(并合,自调匀整)；
- 纤维对齐、伸直；
- 纤维混合；
- 短纤去除；
- 去除异物(包括灰尘)和棉结；

- 加捻;
- 卷绕,清纱(去除疵点),上蜡/润滑(针织纱)。

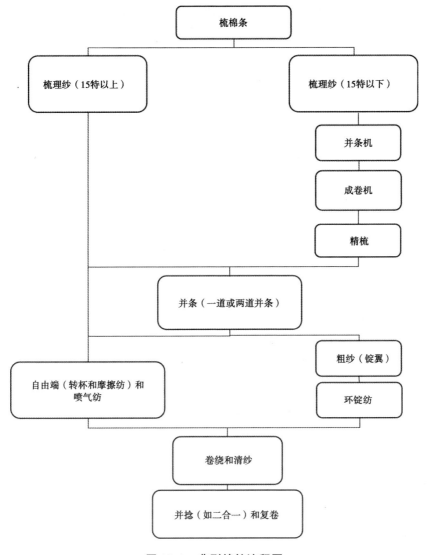

图 19.1　典型棉纺流程图

　　准备工序和纺制的最终目的是以最少浪费且尽可能经济高效方式获得较粗的棉条和纱线。棉条中的纤维是自由随机排列的且纤维混合也并不均匀,棉条中也包含不理想的短纤维、带弯钩的纤维、异性杂质。

　　所制得纱线中纤维尽可能伸直,有序,排列整齐(平行),纱线的外观和成分尽可能均匀,纱线疵点、缺陷、杂质、突出的毛羽和短纤维尽可能少。纱线应制成合适的卷装,以适应后续织物生产过程。

大体上有两种工艺路线,一条是生产粗梳纱的工艺路线,另一条是生产精梳纱的工艺路线,后者涉及额外的精梳工序。(见图 19.1)

Lesson 20   The opening, blending, cleaning, and carding

When cotton and short-staple man-made fibres (MMFs) are delivered to a spinning mill, they are usually received in the form of compressed, high density bales of 1.5m × 0.5m × 0.5m in dimensions, weighing 230kg–250kg, and of 613kg/m^3 in density. A typical production rate for, say, a medium size mill would be of the order of 500kg/h. This means that the equivalent of one bale of fibre would need to be processed every 0.5 hour. Depending on the fineness, length and density of the fibre type to be converted to yarn, the bale can comprise 1.5 to 5 × 1,010 fibres (50 billion); this calculates to approximately 30 million fibres per second removed from the baled stock.

The most practical way of doing this is to remove clumps or tufts of fibres from the bale and then progressively reduce the size of these tufts into smaller tufts or tuftlets ultimately reaching the state of a collection of individual fibres which can be subsequently spun to make the required yarn. Therefore, in preparing materials for spinning, the primary purpose of blowroom operations is to convert the baled fibre mass into the individual fibre state and assemble the fibres in a suitable form for subsequent processing by the intermediate processing stages to spinning. The individual fibre state is essential because these post-blowroom stages require the material to be a linear mass of disentangled fibres, so that fibres can be made to slide past each other in order to uniformly reduce the linear density of the mass.

During this action fibre friction straightens and parallelises the fibres in the cross-section of the reduced linear mass prior to it being twisted to make yarn. The action of breaking the baled fibre mass down into initially large and then smaller size tufts (tuflets) is termed opening (fibre opening) and the converting of small size tufts into individual fibres is called carding. With natural fibres such as cotton, the baled fibre mass will contain impurities, such as leaf, seed, trash and dust particles. It is essential that as much as possible of the impurities are removed so as to produce yarns of high quality. Although certain post-blowroom processes can remove impurities, the opening and carding actions of the blowroom machines enable most of the impurities to be extracted at these early stages. We therefore refer to the term cleaning as the removal of impurities from fibre tufts during opening and carding.

When the baled fibre mass is to be opened into small tufts, a sequence of machines is used to perform this task and as they can also carry out cleaning of the tufts, the machine sequence is called an opening and cleaning line; machines making up such a line are generally referred to as opening and cleaning machines. In the progressive opening of the fibre mass, tufts from one machine are fed to the next in the series for further opening. This is achieved by airflow through pipe ducting linking the machines and is described as pneumatic transport. The tufts are effectively blown from one machine to the next in line, hence the term "the blowroom".

As the tufts arrive at a machine, they are collected in a bin or hopper to form a new assembled mass of tufts which is then worked on by the machine. The opportunity therefore arises for tufts to be mixed or blended as they are being collected to form the assembled mass. Thus, tufts from different bales and different parts of a bale can be blended. This is an essential part of the blowroom process, since fibres at different regions of a bale will have noticeable differences in their properties, particularly natural fibres like cotton in which maturity, length, strength and elongation may differ. Unless tufts are well blended, the difference in fibre properties can result in poor processing performance upstream from the blowroom, such as high end breakage rates in spinning, and in a lower quality of the resultant yarn, e.g., lower yarn strength and evenness.

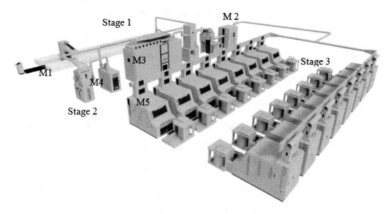

Figure 20.1   Blowroom: opening, cleaning, blending and carding installation

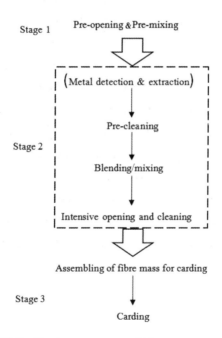

Figure 20.2   Basic sequence of blowroom operations

We speak therefore of blowroom blending as the mixing of opened fibrous tufts to produce a homogeneous mass which facilitates consistent yarn quality. The word "facilitate" is important here, because other process factors can also cause lower yarn quality and reduced production efficiency.

Once the fibre mass is suitably opened, cleaned and blended it arrives at the carding machine and here the tufts are descretised into individual fibres which are reassembled into the form of a twistless rope of disentangled fibres, i.e., a linear mass of fibres held together largely by interfibre friction. This form of fibre mass is called a carded sliver and is coiled into large cans (card sliver cans) ready for the upstream processes. At one time the carding process was housed separately from the earlier opening and cleaning machines. Then the fibre mass was fed in the form of laps to the card. Modern mills now tend to have these machines in the same location with pneumatic transport of tufts to the cards. Fig. 20.1 shows a modern blowroom, and Fig. 20.2 shows a flow chart indicating the function of each stage.

Stage 1 uses a special machine (M1) for the automatic removal of tufts from lines of bales ("bale lay downs"), Stage 2 involves the opening, cleaning and blending machines (M2, M3, M4) and Stage 3 involves the carding machines (M5). As shown, the opening, cleaning and blending line feeds sixteen cards, which indicates that the output from Stages 1 and 2 is much greater than the production capacity of one card. A machine throughput balance must therefore be established when planning and designing a blowroom operation. From this overview we can now consider the basic principles involved at each stage and the production calculations that can be used to determine the appropriate number of cards for a blowroom line of a given production output.

生词与词组

1.blowroom 清花室

2.short-staple man-made fibre 人造短纤

3.tuft [tʌft]*n.* 丛；撮

4.tuftlet [tʌftlɪt]*n.* 丛生；小簇；小束

5.post-blowroom 清花后

6.disentangle [ˌdɪsɪnˈtæŋgl]*v.* 解脱；解开（结扣等）

7.opening [ˈəʊpənɪŋ]*n.* 开松

8.carding [ˈkɑ:dɪŋ]*n.* 梳理

9.impurity [ɪmˈpjʊərɪti]*n.* 杂质

10.leaf [li:f]*n.* 叶子；茎

11.pneumatic [njuːˈmætɪk]*adj.* 气动的

12.hopper [ˈhɒpə(r)]*n.* 储料斗；棉箱

13.upstream [ˌʌpˈstriːm]*adj.* 上游的

14.homogeneous [ˌhɒməˈdʒiːniəs]*adj.* 同种类的

15.reassemble [ˌriːəˈsembl]*v.* 重装；聚集

16.coiled [kɔɪld]*adj.* 盘绕的；绕成圈的

17.mill [mɪl]*n.* 工厂

18.flow chart 流程图

译文

第 20 课　开棉、混棉、清棉和梳理

当棉花和人造短纤维（MMF）被运送到纺纱厂时，它们通常以尺寸为 1.5 米 × 0.5 米 × 0.5 米、重量为 230-250 千克、密度为 613 千克 / 立方米的高密度压缩棉包的形式被接收。按典型的生产率，一家中型棉纺厂生产速度约为 500 千克 / 时。这意味着每半小时需要处理一包纤维。按纤维的细度、长度和密度换算为纱线，一包含有 1.5 至 5 × 1010 根纤维（500 亿）；计算得出每秒从包中带走约 3000 万根纤维。

常用方法是从包中带走纤维团块或簇绒，然后进一步将这些簇绒减小为尺寸更小的簇绒或纤维束直至成为单纤维的集合体，再将其纺成所需的纱线。因此，在纺纱材料准备阶段，开清工序的主要目的是将成包的纤维块转化为单独的纤维状态，并将纤维集合成适于中间阶段至纺纱阶段加工的形态。单独的纤维状态是必不可少的，因为开清后的工序要求材料是松解的线性纤维束，这样可使纤维滑过彼此形成均匀的低线密度束。

在此动作期间，纤维摩擦伸直，平行，线性束横截面减少，再被加捻以制成纱线。将块状的纤维团分解成先大后小的簇绒（簇）的动作称为开松（纤维开松），而将小簇绒转化为单根纤维称为梳理。对于棉花等天然纤维，成包的纤维团含有杂质，如树叶、草籽、垃圾和灰尘颗粒。为生

产出高质量纱线,尽可能地除杂是至关重要的。尽管一些开清后道工序也可去除杂质,但开清机的开松和梳理动作可先去除大部分杂质。因此,我们将在开松和梳理过程中从纤维簇中去除杂质称为"清理"。

由一系列机械来完成成包的纤维团开松成小簇的任务,它们也可进行簇的清理,该机器组称为开松和清理生产线;组成这条生产线的机器通常被称为开松机和清理机。在纤维团的渐进开松过程中,来自一台机器的簇绒被送入机组中的下一台机器以进一步开松。这由气流穿过管道连接的机器来实现,被称为气力传输。棉纤维团被有效地从一台设备吹送到下一台设备,因此有了"清花"一词。

当簇绒到达一台机器时,它们被收集在箱或料斗中以形成新的簇绒集合团,再由机器加工。在簇绒被收集成集合团前,这是簇绒进行混合或掺和的机会。因此,来自不同包或包内不同部位的簇绒可进行混合。这是开清流程的重要组成部分,因来自不同区域棉包的纤维在特性上会有显著差异,尤其是像棉花这类天然纤维,其成熟度、长度、强度和伸长率可能不同。除非簇绒被很好地混合,否则纤维性能的差异会导致开清后续的加工性能不佳,如纺纱的断头率高及成纱的质量低,如纱线强度和均匀度较低。

因此,我们将混合开松的纤维簇,以产生均匀的团块促进纱线质量的一致称为开清混合。"促进"这个词在这里很重要,因为其他工艺因素也会导致纱线质量下降和生产效率降低。

纤维团被适当地开松、清洁和混合后进入梳理机,此处的簇绒被分解成单纤维,松解的纤维再重新聚合成无捻绳带形式,即由纤维间摩擦维持的线性纤维团。这种形式的纤维团称为"梳理条"(生条),并被圈绕在大筒内(梳理条筒)为后续工序做准备。以往,梳理过程与前期的开松和清理机器分开进行,再将纤维团以成卷的形式送入梳理机。当前,现代工厂已将这些机器安在同一位置,由气力将簇绒输送至梳理机。图20.1为现代清花工序,图20.2为各工序的功能流程。

第一个阶段是使用特定设备(M1)从棉包线("棉包平铺")上自动抓取纤维团;第二个阶段涉及开松、清理和混合设备(M2、M3、M4),第三个阶段涉及梳棉机(M5)。如图所示,一条开松、清理和混合生产线喂入十六台梳棉机,这说明第一个阶段和第二个阶段的产量远远大于一台梳棉机的生产产量。因此,在规划设计清花车间时,设备的产出平衡必须确立。从这个概述中,我们现在可以考虑每个阶段所涉及的基本原则以及用于确定给定产量的清花线的适当梳理数量的生产计算。

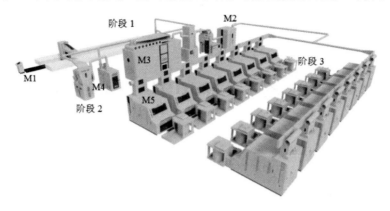

图 20.1　清花车间：开松、清理、混合和梳理设备

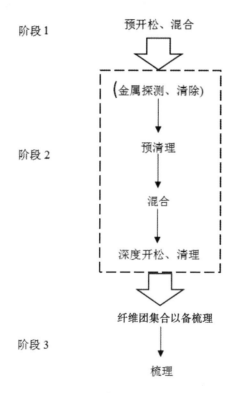

图 20.2　清花车间操作的基本工序

Lesson 21 Spinning process

1.Drawing

Drawing, on a drawframe, represents the first process (called a drawframe passage) applied to the card sliver (containing some 30,000 fibres in its cross-section), with the intention to reduce (attenuate) the sliver linear density until the desired linear density for spinning (some 100 fibres in the yarn cross-section) is achieved. The drawframe operation can also be either linked to, or integrated with, the card, with coiler delivery speeds approaching 500m/min being possible in such cases, higher drafts (2 to 2.5) improving fibre orientation (alignment). The process of sliver attenuation is called drafting. Commonly two drawframe passages (single or twin delivery) occur between carding and spinning. Drawing therefore involves the processes of drafting and doubling, with good fibre control being of the essence throughout. Doubling refers to the action of combining two or more slivers during a process, such as drawing, doubling taking place at the input to the drawing stage. Lateral fibre blending and evening (autolevelling) also take place during the drawing process. Individual drives and sensors enable "self-learning" and "self-adjusting" drawing processes, with expert systems facilitating the detection of faults and the optimum setting and operation of drawframes.

2.Drafting

It is possible to spin yarn directly from either a drawframe or card sliver, and this is in fact accomplished in certain high draft sliver spinning systems (drafts above 120) and often in rotor spinning. Nevertheless, it is difficult to produce even ring-spun yarn, particularly fine yarn, in such a way, the reasons being that it requires a very uniform input sliver and a very precise control of the feed sliver as well as very precise drafting and fibre control, since very high drafts are involved and the beneficial effects of sliver feed reversal and doubling are eliminated.

Good drafting is difficult to achieve under such high draft conditions, there generally being an optimum draft, and the total draft necessary to achieve the required sliver and yarn linear densities is normally accomplished by drafting in stages. The sequence of processes, called drawing, is required to gradually and in a controlled manner, through a process of drafting, reduce the sliver linear density while controlling the movement and alignment of the fibres. Drafting takes place by:
  • Fibre straightening (decrimping);
  • Fibre elongation;
  • Fibre sliding (relative movement).

3.Combing

Combing is used when high-quality fine cotton yarns (finer than approximately 15tex) are required, improving the fibre straightening and alignment and removing short fibres, fibre hooks and any remaining neps and trash particles, thereby enabling finer, stronger, smoother and more uniform yarn to be produced. Usually one (sometimes two) drawframe passages are used prior to combing so as to straighten and orientate the fibre hooks, thereby enabling optimum combing performance. The waste material removed during combing is referred to as noil (or sometimes as comber waste). The noil percentage normally falls between about 5% and 15%.

In preparation for combing, a number of drawframe slivers are combined in a "lap winder" to produce a comber lap consisting of a closely spaced sheet of slivers wound onto a cylindrical holder. The speed of the lap winder exceeds 100m/min, with a draft between two and four. The high doubling combined with the low draft result in a considerable improvement in the evenness of the lap. The correct tension is also required during the winding of the lap. Very low tension results in a soft lap, requiring more storage space and which is also prone to damage during subsequent handling and transport, while very high tension makes it difficult to unwind the lap at the comber, particularly the last few layers, sometimes leading to "split laps". The lap forms the input to the comber. Essentially the lap is combed and drafted into fibre webs, which are layered at the comber table to form a sandwich, which is drafted to form the combed sliver, the latter being coiled into a can ready for the next stage, called "finisher" drawing, using a conventional drawframe, with single or twin delivery and autolevelling. The combing machine can have some eight combing heads. Comb production rates up to 70kg/h, and nip speeds up to 450 per min are not uncommon, with automatic lap change and transport and batt piecing and a self-cleaning top comb.

4.Roving

For carded yarns, typically two drawframe passages precede the roving operation, while for combed yarn, one drawframe passage precedes combing and one succeeds combing. Roving production, on a machine termed a speed frame or flyer, is the final process prior to ring-spinning. It is popularly carried out using an apron drafting system and then inserting a low level of twist into the roving in order to impart sufficient cohesion and condensing to the roving to facilitate uniform and controlled drafting during ring-spinning, particularly during the pre-drafting stage.

Simple roller drafting was used in some systems up to 70 years ago but due to its relatively poor fibre control, it was replaced by the apron drafting system which provides far better fibre control, due to the aprons exerting a very light pressure on the fibres, until they reach the nip of the front rollers. This has become widely adopted as the drafting system in both the roving and ring-spinning operations. Drafts range between 3 and 16, with the roving linear density typically between 300tex and 600tex, the drafted twisted strand is wound onto a bobbin using a flyer and bobbin arrangement, one turn of twist being inserted with each rotation of the flyer, the latter also protecting the roving from balloon formation and air currents. The bobbin has a higher surface speed than the flyer which winds the twisted roving onto the bobbin. Automatic (integrated) bobbin doffing, followed by bobbin loading and transport, is now a reality. Flyer speeds of over 1,500revs/min are possible.

生词与词组

1.draw frame 并条机

2.drafting ['drɑ:ftɪŋ]*n.* 牵伸

3.self-learning [self ˈlɜ:nɪŋ]*n.* 自主学习；自学习

4.self-adjusting [self əˈdʒʌstɪŋ]*n.* 自调节能力；自调整

5.rotor spinning 转杯纺纱；气流纺纱；气流纺

6.ring-spun yarn 环锭纱

7.sliding [ˈslaɪdɪŋ]*adj.* 滑动的；滑移的

8.noil [nɔɪl]*n.* 落棉

9.lap winder 成卷机；成卷装置

10.comber [ˈkəʊmə]*n.* 精梳机；梳棉机

11.cylindrical holder 圆柱形支架

12.prone [prəʊn]*adj.* 有……倾向的；易于……的

13.split lap 分圈

14.batt piecing 棉絮搭接

15.self-cleaning [self ˈkliːnɪŋ]*n.* 自清洁

16.roving [ˈrəʊvɪŋ]*n.* 粗纱

17.apron [ˈeɪprən]*n.* 皮圈

18.cohesion and condensing 黏聚与凝结

19.pre-drafting [pˈriːdrˈɑːftɪŋ]*n.* 预牵伸

20.balloon [bəˈluːn]*n.* 气圈；气流圈

译文

第 21 课　纺纱工序

1. 并条

并条机上的并条指使用梳理条（其横截面包含约 30000 根纤维）的第一个过程（称为并条机通道），其目的是减少（减弱）条子线密度，直至达到纺纱所需的线密度（纱线横截面约有 100 根纤维）。并条机操作既可与梳棉机相连，也可整合在一起。此条件下，其卷取输送速度接近 500 米/分，其较高的牵伸倍数（2 至 2.5）可改善纤维取向（对齐）。条子变细的过程称为牵伸。在梳棉机和纺纱之间通常有两条并条机通道（单眼或双眼）。因此，并条涉及牵伸和并合两个过程，良好的纤维控制至关重要。并合是指在并条机的输入端将两根或两根以上的条子进行并合。并条期间，后续纤维需进行混合、均匀（自调匀整）。独立的驱动器和传感器可实现并条过程"自学习"和"自调整"，专家系统有助于检测故障以及并条机合理的设置、操控。

2. 牵伸

来自并条机或梳棉机的条子直接纺纱，这在一些大牵伸的条子纺纱系统（120 倍以上的牵伸）中已可实现，通常为转杯纺纱。然而，环锭纱，特别是细纱，很难以这种方式生产，原因是它需要喂入的条子非常均匀且可非常精确地控制喂入条子，以及非常精准地牵伸和控制纤维，因相关的牵伸倍数高且可消除条子反向喂入、并合的影响。

高牵伸很难实现良好的牵伸，通常需一个最佳牵伸倍数，条子和纱线达到预定线密度的总牵伸通常由分阶段牵伸来完成。这一系列过程称为并条,需要通过牵伸以渐进式的方式降低条子线密度,同时控制纤维的移动和排列。牵伸过程作用有:

- 纤维伸直(消除卷曲);
- 纤维伸长;
- 纤维滑移(相对运动)。

3. 精梳

当需要高质量的细棉纱时(细度小于 15 特),需使用精梳来改善纤维的伸直平行度,去除短纤维、剩余棉结和杂质,以生产更细、强力更高、光泽更好、条干更好的纱线。为了伸直、定向纤维弯钩,通常在精梳前使用一道(有时是两道)并条通道,从而获得最佳的精梳效果。在精梳工序中除去的废料称为落棉(或有时称为精梳废料),落棉率通常在5%-15% 之间。

在精梳准备工序中,许多条子在"绕卷机"上合并成紧密条片,绕在圆柱形支架上形成棉卷。绕卷机的速度达 100 米 / 分,牵伸倍数为 2-4,高倍速结合低牵伸可显著改善棉卷的均匀度。绕卷需要恰当的张力,张力太低会导致棉卷柔软,需要更多的存储空间,且在后续的处理和运输过程中也易损坏,而张力太高则在精梳机上难以退卷,尤其是最后几层,有时会引起"分圈"。成卷喂入精梳机。基本上,棉卷经过精梳和牵伸成纤维网,再铺在精梳台形成夹层,经牵伸制成精梳条,再被圈绕在筒内供后续使用,此称为"末道"并条,常规并条机具有单或双通道和自动匀整。精梳机有 8 个精梳位。精梳产量高达 70 千克 / 时,钳口速度达 450 次 / 分的精梳机较为少见,且带有自动棉卷更换和输送以及棉絮接头和自清洁顶梳。

4. 粗纱

通常普梳纱在粗纱操作之前有两道并条,而精梳纱在精梳前后各设一道并条。粗纱生产,在称为速纺机或锭翼的机器上,是环锭纺之前的最后一道工序,通常使用皮圈牵伸系统,然后在粗纱中加入少量捻度以赋予足够的内聚力,凝聚成粗纱以促进在环锭纺纱过程中均匀和受控的牵伸,特别是在预牵伸阶段。

70 年前,就已开始使用的简易罗拉牵伸系统,由于其纤维控制力相对较差,现在已由皮圈牵伸系统所取代,由皮圈对牵伸区内直至前罗拉钳口的纤维施加了小压力可更好地控制纤维。在粗纱和环锭纺细纱工序中,这种牵伸系统已被广泛使用。一般,粗纱牵伸倍数为3–16,线密度为300–600特,使用锭翼和成排筒管将牵伸加捻的线缠绕到筒管上,锭翼每旋转一圈加入一圈捻度,后者还可保护粗纱免受筒管成形和气流的影响。筒管的表面速度比锭翼高,可将加捻粗纱缠绕到筒管上。自动(集成)筒子落纱,接着筒子装载和传输,已可实现。锭翼转速达1500转/分。

Lesson 22　Spinning system

1.Introduction

Spinning can be divided into the following three basic operations:

• Attenuation (drafting) of the roving or sliver to the required linear density;

• Imparting cohesion to the fibrous strand, usually by twist insertion;

• Winding the yarn onto an appropriate package.

Spinning systems presently employed include ring (including the compact and "two-strand" systems), rotor (open-end), self-twist, friction (also open end), air jet, twistless and wrap-spinning, with the first mentioned two systems by far the most important for cotton spinning, together accounting for far over 90% of the cotton yarn produced globally. One rotor spindle is generally taken to equal five or more ring spindles in terms of yarn production capacity. Ring-spinning accounts for some 70% of global long and short staple yarn production, rotor spinning for some 23% and air jet vortex for some 3%. The main reason for the dominance of ring-spinning (which is well over 100 years old) over other spinning systems is the superior quality, notably strength and evenness, of ring-spun yarns over those produced by other systems. Very fine ring yarns can also be spun (even as fine as 2tex), the spinning limits being about 35 fibres in the yarn cross-section for combed yarns and 75 fibres in the cross-section for carded yarns. Second in importance, and increasing its share of cotton yarn spinning, is the rotor (open-end) spinning system.

2.Ring spinning

Because of its versatility in terms of yarn linear density and fibre type and also the superior quality and character of the yarn it produces, as a result of good fibre control, orientation and alignment (extent) during spinning and in the yarn, ring spinning (Fig. 22.1) remains by far the most popular system for spinning, particularly for fine yarns. Its main disadvantage is the yarn production rate due to limitations in spindle speed (productivity), due to high power consumption, traveller wear, heat generation and yarn tension.

It is necessary to rotate the yarn package (tube, bobbin), approximately once for each turn of twist inserted, this consuming a great deal of power (approximately 74% required to overcome "skin friction" drag and 25% to overcome yarn wind-on tension), even for small packages. Smaller rings, higher spindle speeds, automatic doffing, compact spinning, on-line monitoring, linked winding and very hard rings and travellers (e.g., ceramic) all play a role in ring spinning maintaining its popularity. About 85% of the total power requirements of a ring-frame is consumed in driving the spindles (depending on yarn density, package size, spindle speed, etc.) the balance being consumed by the drafting and other mechanisms.

3.Two-strand spinning(twin-spun)

Two-strand spinning, also referred to as spin-twist or double-rove spinning, involves two rovings being fed separately to the same double apron drafting system, each strand receiving some twist before they are combined at the convergence point after the front rollers. The Sirospun system uses, for example, a mechanical break-out device and automatic splicing to prevent spinning when one strand breaks. It is also possible to include a filament (flat, stretch or textured).

Spinning limits are about 35 fibres per strand cross-section. With the EliTwist® system (Fig. 22.2), yarns with resultant linear densities from R60 tex/2 (Ne 20/2) to R8 tex/2 (Ne 140/2) can be spun.

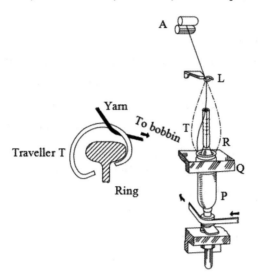

Figure 22.1   The ringframe

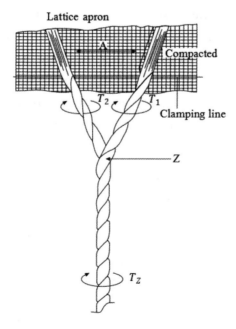

Figure 22.2   Two-strand spinning EliTwist® system

4.Compact (condensed) and related spinning systems

Following upon the two-strand spinning developments, further work was undertaken to produce ring-spun single yarns with superior properties (notably tensile, hairiness, abrasion and pilling). Considerable success has been achieved and compact spinning systems were commercially introduced in the 1990s, in some cases enabling combed yarns to be replaced by carded yarns and two-ply yarns by single yarns. Compact spinning has caused a revival in ring spinning, and compact yarns fetch a premium price. Examples of compact/condensed spinning systems include the EliTe spinning system of Suessen, ComforSpin® (Com4®) of Rieter, CompACT3 of Zinser and RoCoS of Rotorcraft.

5.Bicomponent spinning

Bicomponent yarns, also referred to as bound yarns, have a niche market, these generally combining pre-spun continuous filament yarns (sometimes even staple yarns) with staple fibres to provide improved properties, such as stretch (e.g., Lycra), abrasion and strength. The filament yarn can either be covered by the cotton sheath (i.e., be in the core of the yarn) or be wrapped around the outside of the yarn, the former being more common.

Bicomponent spinning normally involves twisting together either a filament (sometimes water soluble) and a conventionally drafted staple (cotton) strand during the spinning operation, and is particularly attractive for the cost-effective production of superior yarns which can, for example, be woven or knitted without any further operations (i.e., eliminating plying, sizing and steaming). It also enables coarser fibres to be spun into finer yarns, reduces spinning end breakages, allows higher winding speeds and enables yarn and fabric properties to be engineered by suitable selection of the two components and the way in which they are combined.

6.Rotor (open-end) spinning

Rotor spinning, generally referred to as open-end (OE) spinning, since there is a definite break (discontinuity or open end) in the fibre flow prior to yarn formation, was commercially introduced during the late 1960s and is second only to ring spinning in terms of short staple yarn production. It has reached the stage where it is now classed, together with ring-spinning, as "conventional" spinning.

It can produce yarn within the range of approximately 10tex to 600tex, more often 15tex to 100tex, with delivery speeds as high as 300m/min or even higher, although its economics become less favourable at the finer end of the scale. The main disadvantages of rotor-spun yarns compared with ring-spun yarns are their lower strength and the presence of wrapper fibres which adversely affect their handle. The spinning limits are also lower, generally taken to be between 100 and 120 fibres in the yarn.

7.Air jet spinning

Air jet spinning, also termed Vortex spinning, introduced in the 1980s, produces yarn typically in linear densities from about 5tex to 40tex, at speeds up to 400m/min or even higher. The processes prior to air jet spinning are similar to those used for rotor-spinning, although combing is more often used in preparation since dust and trash can obstruct the spinning jets. The yarn more closely resembles, but is weaker than, ring-spun yarns. Air storage (accumulator) systems are not suitable at such high speeds and the yarn storage system of the machine is similar to that of a weaving machine yarn storage feeder, which is self threading, and constantly stores and discharges yarn.

8.Friction spinning

Although cotton, either in 100% or in blends with other fibres, is spun on the friction spinning system, the percentage is very low. The reason for this is largely to be found in the lower quality, strength in particular, of friction spun cotton yarns compared with ring- and rotor-spun yarns due to the lower efficiency of the yarn structure (relative low orientation, extent and packing of fibres), as well as in the fact that the fibre feed speed is too high and yarn tensile force too low for fibre binding at the yarn end. Friction spun yarn are weaker and more twist lively than other yarns, although plying can reduce these drawbacks. The spinning limits for friction spinning cotton are around 100 to 120 fibres in the yarn cross-section.

生词与词组

1.traveller [ˈtrævələ(r)]*n.* 钢丝圈

2.cop [kɒp]*n.* 纱管；纬管

3.ring [rɪŋ]*n.* 锭子

4.compact spinning 紧密纺

5.condensed spinning 集聚纺纱

6.air storage (accumulator) system 气流储存（储能器）系统

7. "two- strand" system 双股系统；双线系统

8.open end 自由端

9.air jet spinning 喷气纺纱

10.wrap-spinning 包缠纺纱

11.spindle [ˈspɪndl]*n.* 锭子

12.vortex [ˈvɔ:teks]*n.* 涡流；涡旋

13.tube [tju:b]*n.* 管；筒

14.bobbin [ˈbɒbɪn]*n.* 筒管

15.skin friction drag 表面摩擦力

16.wind-on tension 卷绕张力

17.on-line monitoring 在线监测

18.Sirospun [saɪˈrɒspʌn]*n.* 赛络纺；赛络纺纱

19.two-ply yarn 双股线

20.bicomponent yarn 双组分纱

21.staple [ˈsteɪpl]*n.* 短纤

22.sizing [ˈsaɪzɪŋ]*n.* 浆纱；上浆

23.steaming [ˈstiːmɪŋ]*n.* 水蒸气处理；蒸化

24.obstruct [əbˈstrʌkt]*v.* 阻隔；阻塞

25.feeder [ˈfiːdə(r)]*n.* 进料器；给料

26.drawback [ˈdrɔːbæk]*n.* 退回

译文

第 22 课　纺纱系统

1. 概述

纺纱包括以下三部分基本操作：

（1）将粗纱或条拉细（牵伸）成所需要的线密度；

（2）通过加捻赋予纱线强力；

（3）将纱线卷绕在合适卷装上。

目前使用的纺纱系统包括环锭纺（包括紧密纺和"双股"赛络纺）、转杯纺（自由端纺纱）、自捻纺、摩擦纺（也是"自由端纺纱"）、喷气纺、无捻纺和包缠纺，首先提及的两个系统对棉纺至关重要，它们共同占全球棉纱的90%以上产量。就产量而言，一般认为一个转杯纺锭相当于五个或以上的环锭纺锭子。环锭纺纱约占全球长短纤纱产量的70%，转杯纺约占23%，喷气涡流纺纱约占3%。环锭纺（环锭纺已有100多年的历史）比其他纺纱系统占优势的主要原因是环锭纺纱线质量优于其他纺纱技术，特别是强度和条干。环锭纺也可纺出非常细（甚至细至2特）的纱线，纺纱极限是精梳纱横截面35根纤维、粗梳纱横截面75根纤维。第二重要的是转杯（自由端）纺纱，它在棉纱纺纱中所占的份额正在增加。

2. 环锭纺

因纱线线密度和纤维类型的多样性,以及其所制纱线的品质和性能卓越,环锭纺的纺纱过程和纱线内的纤维控制、取向和排列(程度)良好,其(图 22.1)仍然是最流行的纺纱系统,特别是细纱。环锭纺最主要的缺点是高耗能、钢丝圈磨损、发热和纱线张力,锭子速度(生产效率)受限。

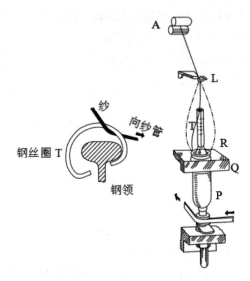

图 22.1　环锭纺纱机

纱线绕成卷装(管纱、筒管、纱管),每转一转加入一圈捻度,这会消耗大量动力(大约 74% 来克服"表面摩擦"阻力,25% 克服纱线卷绕张力),即使是小包装。更小的钢领、更高的锭速、自动落纱、紧密纺、在线监测、连接络筒和非常坚硬的钢领和钢丝圈(例如陶瓷)都对环锭纺保持流行起到关键作用。在环锭纺设备中,锭子转动(依据纱线密度、筒装尺寸、锭子速度等)所消耗的能量约占总能耗的 85%,剩下由牵伸和其他机构耗能。

3. 双股纺(双纺)

双股纺纱,也称为捻纺或双粗纱纺纱,是将两根粗纱分别喂入同一个双皮圈牵伸系统,每根粗纱在前罗拉之后的汇合点合并之前都加一些捻度。例如,赛络纺技术使用了机械断头装置和自动接头装置,以防止一根粗纱断裂后继续纺纱。它也可包含一根长丝（普通丝、弹力丝或变形丝）。纺纱极限大约为纱线横截面 35 根纤维。使用 EliTwist® 系统（图 22.2），可以纺出线密度为 R60 特 /2（英支 20/2）到 R8 特 /2（英支 140/2）的纱线。

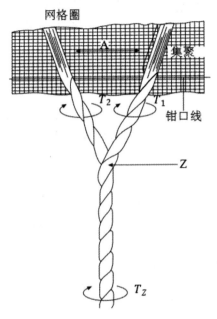

图 22.2　双股纺纱 EliTwist® 系统

4. 紧密纺及相关纺纱系统

紧跟双股纺纱技术而发展的是超性能（显著强力、毛羽、磨损和起球性能）的环锭单纱。已经取得了相当大的成功，紧密纺系统在 20 世纪 90 年代投入商业使用，一定程度上精梳纱可以被普梳纱取代，两股纱可以被单纱取代。紧密纺推动了环锭纺的复兴，紧密纺纱线的价格也更为高昂。紧密 / 集聚纺纱技术应用范例包括绪森的 EliTe 纺纱系统，立达的 ComforSpin®（Com4®），青泽的 CompACT3 和罗卡斯的 Rotorcraft。

5. 双组分纺纱

双组分纱线，也称为结合纱线，市场前景较好，它们通常将预纺连续长丝纱线（有时甚至是短纤维纱线）与短纤维相结合，以改进纱线性能，如弹性（莱卡）、耐磨性和强度。长丝纱线可以被棉包覆（即在纱线的芯部）或缠绕在纱线的外面，前者更常见。

6. 转杯纺（自由端纺纱）

转杯纺，通常被称为自由端（OE）纺，因在纱线形成之前纤维流中存在一定的断裂（不连续或开放端）。20 世纪 60 年代后期被商业引入，其短纤纱的产量仅次于环锭纺。它已经到了与环锭纺一同归类为"传统"纺纱的阶段。

它可生产 10-600 特的纱线，常见的是 15-100 特，生产速度达 300 米/分以上，在高支段纱线的经济效益已欠佳。与环锭纺纱相比，转杯纱的主要缺点是强度较低、带有缠绕纤维，其对手感有不利影响。纺纱极限更低，一般认为纱内包含 100-120 根纤维。

7. 喷气纺纱

喷气纺，也称为涡流纺，20 世纪 80 年代推出，通常生产线密度为 5-40 特的纱线，其速度高达 400 米/分以上。喷气纺的前道流程与转杯纺相似，因灰尘和杂质会堵塞纺纱喷嘴，准备阶段需设精梳。这种纱线更类似于环锭纱，但比环锭纱弱。储气（蓄能器）系统不适用于如此高的速度，机器的储纱系统类似于织机储纱器，可自动生头，不断储存和排出纱线。

8. 摩擦纺纱

无论是 100% 棉还是棉与其他纤维的混纺，都可在摩擦纺纱系统上进行纺制，但比例非常低。其原因主要在于摩擦纺棉纱与环锭纺和转杯纱相比质量较低，尤其是强度较低，这归因于其成纱结构的效率较低（相对较低的取向性、延伸性和包缠纤维），以及纤维喂入速度太高、纱线张力太低而无法在纱线端部接结纤维。尽管合股可减少这些不足，但摩擦纺纱比其他纱线更弱，更易捻缩。 摩擦纺纱的纺纱极限是纱线横截面含有 100-120 根纤维。

Lesson 23   Twisting, winding and clearing

1.Twisting

Yarn twisting is normally applied to improve yarn evenness (CV), strength, extension and abrasion resistance and to reduce twist liveliness by balanced twist, hairiness and fibre shedding (fly) and to produce speciality (fancy) yarns. Balanced twist is normally achieved when the plying twist is approximately two-thirds that of the single yarn twist, and in the opposite direction.

The twisting operation, also referred to as plying, is the process whereby two (sometimes more) yarns are twisted to form a two-ply (or multi-ply) yarn. Traditionally, this was done on a ring frame (ring twister) but today it is almost exclusively carried out on a two-for-one twisting machine, three-for-one twisting systems having also been developed. Assembly winding is used to assemble two ends of yarn on one package in preparation for two-for-one twisting. It is particularly important to ensure that the two yarns are wound at the same tension. The assembly wound package remains stationary, the yarn passing through a guide mounted on a rotating arm which can freely rotate, through the hollow rotating spindle, then through an eyelet (outlet hole) and from there via a yarn guide and yarn take-up rollers to the yarn winding head. One revolution of the spindle inserts one turn of twist into the yarn while the rotating eyelet simultaneously inserts a turn of twist in the yarn in the balloon. Thus two turns are inserted per spindle revolution. Spindle speeds as high as 13,500revs/min and delivery speeds up to 60m/min are possible, the unit can be with or without balloon control.

2.Winding and clearing

Winding, re-winding as it is sometimes called, is aimed at transferring the yarn from the spinning packages (referred to as tubes, bobbins), which normally hold relatively short lengths of yarn, into packages (cones, cheeses, etc.) which can hold considerably longer lengths of yarn more suitable for the subsequent processes, such as yarn preparation, weaving, knitting, package dyeing, etc. The winding process also provides an opportunity for unwanted yarn faults (e.g., slubs and thin or weak places, foreign fibres, trash, etc.) to be classified and removed (i.e., yarn clearing) and the yarn to be lubricated. The latter is often referred to as waxing in the case of knitting since it entails the use of a solid wax disc for lubricating the yarn. Clearers may be either of the capacitance or optical types or even a combination of these.

生词与词组

1.twist liveliness (torque) 捻度不稳定（扭矩）
2.ring frame (ring twister) 环锭纺纱机（环锭捻纱机）
3.two-for-one twisting machine 二合一捻线机；倍捻机
4.eyelet ['aɪlət]*n.* 孔；小孔；眼孔
5.mounted ['maʊntɪd]*adj.* 安装好的；固定好的
6.revolution [ˌrevəˈluːʃ(ə)n]*n.* 旋转；回转
7.cone [kəʊn]*n.* 锥形；锥体；圆锥
8.cheese ['tʃiːz]*n.* 筒子；筒子纱
9.package dyeing 筒纱染色

10.slub [slʌb]*n.* 粗节；糙粒

11.thin or weak place 薄弱环节

12.foreign fibre 外国纤维

13.wax disc 蜡盘

14.capacitance [kəˈpæsɪtəns]*n.* 电容；电容量

译文

第 23 课　捻合、卷绕和清纱

1. 捻合

　　纱线加捻通常用于提高纱线均匀度（CV）、强度、伸长率、耐磨性，通过平衡捻度、毛羽和纤维脱落（飞花）来降低捻度不稳定性（扭矩），并可制造特种（花式）纱线。当合股捻度约为单纱捻度的三分之二且方向相反时，通常可平衡捻度。

　　加捻操作，也称为捻合，是将两根（有时更多）纱线加捻以形成两股（或多股）纱线的过程。传统上，这是在环锭细纱机（环捻机）上完成的，但如今几乎全在二合一捻线机（倍捻机）上进行，三合一捻线系统也已开发出来了。将两根纱线合并卷绕到一个筒子上，为倍捻机加捻做准备。确保两根纱线以相同的张力卷绕尤为重要。集绕的筒子保持静止，纱线通过安装在可自由旋转的旋转臂上的导纱器，通过中空旋转锭子，然后通过孔眼（出口孔），再通过导纱器由纱线卷取辊绕在纱线卷绕头。锭子旋转一圈，在纱线中加入一圈捻度，而旋转的孔眼同步在气圈内的纱线上加入一圈捻度。因此，每当锭子旋转一周，纱线获得两个捻回。锭子速可达 13500 转 / 分，输送速度可达 60 米 / 分，整个设备可带或不带气圈控制。

　　2. 卷绕与清纱

　　卷绕，有时也称为回绕，旨在将纱线从纱管（称为"纱管、筒管"，这些筒子通常容纳的纱线长度相对较短）中转移到筒子（锥筒、圆筒等，可以容纳相当长的纱线），更适合后续工艺，如纱线准备、编织、针织、卷装染色等。络筒可以将一些不需要的纱线疵点(如粗节、细节、异性纤维、杂质等)分类去除,也可进行润滑。在针织条件下,后者常被称为"上蜡",因为它需要使用固体蜡盘来润滑纱线。清纱器可以是电容式或光电式,也可以是二者的结合。

Lesson 24　The woollen process

1.The terms "woollen" and "worsted"

The basic difference between the two is that in the worsted system all short fibres are removed and the remaining long ones are aligned parallel, while in the woollen system there is no removal of short fibres, so some fibres lie parallel and others randomly. The following from *Textile Terms and Definitions* (10th Edition) describe the differences between the two systems: Some would say that the terms "woollen" and "worsted" have become system descriptive, with "wool" being added to describe content, for example, "wool worsted". Woollen yarns being so rarely 100% wool, a description of the blend is usually used if required, for example, 100% wool woollen spun or 100% wool woollen. Woollen, woollen yarn or woollen fabric is descriptive of the fibre that is wool fibre spun on the woollen system.

Woollen spun, woollen type fabric or condenser spun is descriptive of the system—that is any fibre spun on the woollen system. Worsted, worsted yarn or worsted fabric is descriptive of the fibre—that is wool fibre spun on the worsted system.

2.The woollen process

A woollen fabric (as distinct from a worsted one) is made from yarns comprising of wool fibres of variable length, which have been spun on the condenser or woollen spun system. The fibres are allowed to lie haphazardly in spinning and the resultant yarns have a roughish appearance and full handle. Although the raw material for both woollen and worsted yarns is wool fibre, there are important differences. In woollen spinning a wide range of shorter wool types can be used in varying proportions in a blend, together with a limited amount of reprocessed or re-used wool in order to reduce the cost. In worsted spinning only pure new wool fibres of the longer type are used. Certain man-made fibres such as polyester can be blended in varying proportions with pure new wool and spun on either woollen or worsted systems, but such yarns will be neither "woollen" nor "worsted". The main processes in woollen yarn production are described briefly as follows:

Sorting: This was at one time a highly-skilled manual operation to select and divide the fleece into different qualities. It is now rarely used for that purpose, but occasionally to remove heavily contaminated, matted or weathered wool and heavily stained or pigmented patches.

Scouring: Wool in its raw or greasy state is cleaned by mechanically passing it through a series of scouring bowls containing hot water and detergent, then rinsing and drying. The main contaminates removed during this process are wool grease (lanolin), animal sweat (suint), animal wastes and mineral dirt picked up from the grazing area. Depending on the country of origin, sheep wool type, fibre length and fineness, a minimum of 20% of the greasy wool weight will be lost during scouring. In extreme cases only 20% of the greasy wool weight might be wool fibre. The average Australian wool yields 65% clean, but this figure is slowly rising as farming methods improve.

Carbonising: After scouring some wools contain seeds previously picked up by the sheep and these are removed by carbonising. This is a process that carefully treats the scoured wool with acid, dries it and then crushes the seeds or burrs into a powder that falls from the wool. As carbonising tends to weaken and discolour wool, it is processed as a small percentage of a blend.

Blending: This describes the mixing of different fibre lots, which will provide the required quality and performance characteristics of the end product, at a specific price. Fibre lubricants are added at this time to improve processing performance. Depending on blend and end product, between 2% and 15% oil and anti-static additives may be applied.

Carding and Condensing: The blended wool fibres are disentangled and mixed by passing through a series of large cylinders and rollers clothed with wire teeth. As the fibres pass along the card, spacing between the rollers is reduced, the wire teeth become finer and roller speeds increase. The material is transformed into an even web of fibres which is split lengthways into strands of untwisted slubbing, then wound onto spools in preparation for spinning.

Spinning: Twist is added to the untwisted slubbings to convert them into strong, single yarns on the spinning machine. The mule spinning machine has a complex working action and is now more or less obsolete after the arrival of the more productive ring spinning frame. The mule consists of a carriage that travels backwards and forwards across the floor, drawing out the slubbing to the required thickness of yarn, whilst rotating spindles twist and wind the yarn onto tubes. Ring spinning frames have a higher production rate and larger take-up packages and perform the same functions as the mule, but on a faster and continuous basis.

However, the mule produces a better yarn than the ring frame for a given raw material and quality requirement. Improved production speeds gained from more sophisticated engineering methods and computer control has resulted in a renaissance for mule spinning.

Twisting: The resultant spun yarn can be used in single form, or folded with itself (or other yarns) for increased thickness, strength or effect.

Dyeing: This depends on the type of fabric required. This may be carried out on loose fibre, spun yarn or woven cloth.

生词与词组

1.woollen process 粗纺工艺

2.matted ['mætɪd]*adj.* 无光泽的；亚光的

3.pigmented patch 色素斑块

4.bowl [bəʊl]*n.* 槽

5.detergent [dɪˈtɜːdʒənt]*n.* 洗涤剂；去垢剂

6.rinsing [ˈrɪnsɪŋ]*n.* 漂洗；冲洗

7.mineral dirt 矿物质；无机污泥

8.grazing [ˈgreɪzɪŋ]*n.* 牧场；草场；放牧

9.anti-static additive 防静电添加剂

10.animal sweat 动物脂汗

11.wire teeth 锯齿

12.carriage [ˈkærɪdʒ]*n.* 车；牵引车

13.mule [mjuːl]*n.* 走锭

14.sophisticated [səˈfɪstɪkeɪtɪd]*adj.* 复杂的；精密的；先进的

15.renaissance [rɪˈneɪsns]*n.* 复兴；重新兴起

译文

第24课　粗纺工艺

1. 术语"粗纺"和"精纺"

　　二者的基本区别是在精梳毛纺系统中所有短纤维都被除掉，而保留了平行伸直排列的长纤维。在普梳毛纺系统中短纤维并未除掉，部分纤维平行排列，其他纤维随机排列。《纺织品术语和定义（第10版）》中的以下内容描述了两个系统之间的差异：有人会说术语"粗纺"和"精纺"已成为添加了"羊毛"来系统性地描述，如"羊毛精纺"。粗纺纱很少是100%羊毛，如果需要，通常用于混合型表述，例如，100%羊毛普梳纺或100%羊毛粗纺。毛纺、毛纺纱或毛织物是指羊毛纤维通过粗纺系统纺出来的。

　　毛纺、毛型织物或集聚纺是对该系统的描述，即在粗纺系统上纺任意纤维。精梳、精梳纱或精梳织物即羊毛纤维由精纺系统纺制。

2. 粗纺工艺

粗纺织物(与精纺织物不同)由含有各种长度毛纤维的纱线制成,其由集聚纺或粗纺系统纺成。纤维在纺纱过程中被随意放置,所得纱线外观粗糙,手感丰满。虽然粗纺毛纱和精纺毛纱的原料都是羊毛纤维,但有显著差异。在粗纺中,可混纺不同比例的较短羊毛,为了降低成本可混合有限数量的再加工或再利用羊毛。在精纺系统中,只使用长度较长的纯羊毛纤维。一些人造纤维(如涤纶)可按不同比例与纯羊毛混纺,在粗纺或精纺系统中纺纱,但这样的纱线既不是"粗纺"也不是"精纺"。粗纺纱主要生产工艺简述如下。

分拣:这曾经是选择、区分羊毛不同品质的高技术含量的手工工序。现在很少用于该目的,但偶尔会去除严重污染、无光泽或风化的羊毛以及严重沾色或有色斑块的。

洗毛:采用机械方法将毛坯或油脂状的羊毛经一系列含有热水和洗涤剂的洗槽的清洗,再漂洗和烘干。在这个过程中去除的主要污染物是羊毛油脂(毛脂)、动物汗水(脂汗)、动物粪便和从牧区携带的无机污垢。根据原产国、绵羊羊毛类型、纤维长度和细度的不同,在洗毛过程中至少有20%的含脂毛重损失。在极端情况下,仅有20%的含脂毛重是羊毛纤维。澳大利亚平均净毛产量为65%,随着农业生产方式的改进,这一数字正在缓慢上升。

碳化:经洗毛后,一些羊毛会含有一些以前黏着在羊身上的草籽,这些草籽需进行碳化处理。这过程是用酸精细处理羊毛,烘干,然后冲压碳化的草籽或草刺成粉末,并使其从毛中脱落。因碳化作用易降低羊毛强力、使羊毛褪色,所以这个工序只占混合的一小部分。

混合:这表述了不同产地纤维的混合,以特定价格提供最终产品所需的质量和性能特征。为提高加工性能,可添加纤维润滑剂。根据混合和最终产品,可使用2%–15%的油和抗静电添加剂。

梳理和并条:混合后的羊毛纤维由一系列装有锯齿的大圆滚筒和罗拉进行开松和混合。当纤维过梳理机时,辊子之间的间距减小,锯齿变得更细,辊子速度增加。纤维被转化成均匀的纤维网,再沿纵向分束成无捻粗纱,然后绕在前纺的辊筒上。

纺纱:在细纱机上,无捻的粗纱加捻转化为强的单纱。随着更高生产效率的环锭细纱机的出现,操作复杂的走锭细纱机,现在或多或少已经过时了。走锭由一个在地板上来回移动的小车组成,牵伸粗纱到所需的纱线细度,同时旋转锭子进行加捻,再将纱线卷绕到筒管上。环锭细纱机具有较高的生产速度和更大的卷装,且具有与走锭细纱机相同的功能,但速度更快且连续。然而,在给定的原料和质量要求下,走锭细纱机比环锭细纱机生产出的纱线更好。通过更复杂的工程方法和计算机控制可提高生产速度,使走锭纺纱重新兴起。

加捻:最终成纱以单纱形式;为加粗、增强或提效,也可自身(或与其他纱)进行合股。

染色:根据所需织物类型确定,可以对散纤维、纱线或面料进行染色。

Lesson 25 The worsted yarn

A worsted fabric is an all wool cloth made from yarns produced on the worsted spinning system. This system for producing yarns from staple fibres has many more operational stages than those required for woollen yarn spinning. In worsted yarn spinning the drawing out operation from yarn employs several stages of drafting, together with a combing operation.

This produces a yarn in which the fibres lie as parallel to each other as possible, after removal of the shorter fibres. The resultant yarn has a smooth, slick handle and appearance as well as good strength. Worsted yarn spinning produces lighter and finer yarns and fabrics than woollen yarn spinning from the same fibre micron. Wool can be blended with selected man-made fibres and the resultant yarns combine the desirable properties of the components. For example, in a blend of wool and polyester, the fabric would have the superb handle and drape of wool, plus the easy care properties of the polyester.

The early processes in the manufacture of worsted yarns are basically the same as for woollen yarns, namely blending, scouring and carding. There is however one difference in the blending process. The components in a worsted blend are combined in their greasy state and are usually of a similar quality, unlike a woollen blend, so no special blending is necessary since adequate mixing takes place in subsequent processing. The extra processes in worsted yarn spinning after carding are described briefly as follows:

Preparatory Gilling: The carded slivers are prepared for combing by drawing out a group of them between two pairs of rollers, to straighten the fibres. Between the pairs of rollers are pinned bars known as fallers, which control the fibre during drafting and improve the parallelism of the fibre.

Combing: This process is critical in the production of worsted yarns. Between 20 and 30 slivers are fed into a combing mechanism, which removes most of the short fibres (noils) and further straightens the fibres, making them lie parallel to each other. The combed slivers are thereafter referred to as "tops".

Finisher Gilling: By using further gill stages, the tops are blended and arrive at a specified and uniform linear density. They can then be sold to spinners for drawing and twisting into yarn.

Dyeing: If coloured tops are required, they must be dyed before drawing and spinning, by forcing a dyeing solution through them. After further gilling and combing they are ready to be drawn and spun into yarn.

Drawing: The main objective in the drawing process is to gradually reduce the thickness of the top in three or four stages, to a roving from which yarn is spun. This is done by gill box drawing. The roving frame, the intermediate stage between gilling and spinning, drafts a fine sliver to a thickness which is suitable for the spinning frame and either adds a few turns of twist, or lightly rubs the sliver with a rolling action before winding the fibre onto a large bobbin. The twisting or rubbing action gives the fine fibre assembly some cohesion so that it can be pulled from the bobbin as it feeds into the spinning machine.

Spinning: This is the last processing stage where drafting is used to reduce the thickness of the fibre strand. In worsted spinning the material will be drafted at a ratio of 20, that is, the fibre assembly will be 20 times longer at 20 times thinner when it leaves the delivery rollers than it was when it entered the feed rollers. This is the highest draft the fibre will experience. Gill boxes usually have a draft of about 8 and roving frames 12. It is also much higher than woollen spinning where the draft is often less than 2, the final count being fixed at the card. The final count has some bearing on the spinning draft used, as will the type of fibre used—for example, coarse counts spun from synthetic fibre may be drafted at 35 or more. Once the fibre has been drafted, the strand is then twisted and wound onto a package by the ring and traveller unit.

生词与词组

1.slick handle 滑溜的手感
2.pinned bar 销杆；针棒
3.faller[ˈfɔːlə]n.(走锭纺纱机的)坠杆
4.gill box drawing 针梳机

译文

第 25 课　精纺纱

精纺织物是由精纺系统生产的纱线制成的全羊毛织物。这种用短纤维生产纱线的系统比粗纺所需的操作步骤更多。在精纺毛纱过程中，牵伸过程需要经过几道牵伸阶段，设精梳操作。

该方法生产的纱线在去除短纤维后，纤维会尽可能彼此平行。成纱手感光滑，滑溜，外观好，强度高。使用同一种细度纤维的情况下，精纺纱比粗纺纱生产的纱线和织物更轻，更细。毛纤维可与选定的人造纤维进行混纺，混纺纱线结合了各成分的理想性能。例如，将羊毛和聚酯纤维混纺后，织物将具有羊毛的优良手感和悬垂性，同时还有聚酯纤维易于保养的特性。

精纺毛纱的前段生产工艺与粗纺毛纱基本相同，即混合、洗涤和梳理。然而，在混合过程中有区别。与羊毛混纺不同，精纺混纺中的成分在油脂状态下混合，通常具有相似的质量，因此不需要特殊的混纺，因其在后续加工中进行了充分的混合。精纺毛纱梳理后的额外工序简述如下。

预梳理：梳理的条子由两对罗拉牵引出一组来准备梳理，以拉直纤维。一对罗拉之间设固定杆，被称为压杆(压力棒)，在牵伸过程中控制纤维运动并可提高纤维的平行度。

精梳：精梳是精纺纱线生产的关键工序。将 20 至 30 条棉条送入精梳机除去大部分短纤维（落毛），并进一步伸直纤维使它们彼此平行。精梳过的棉条此后称为"毛条"。

针梳：通过进一步的针梳，毛条混合成特定、均匀的线密度，然后就可以卖给纺纱厂进行并条加捻成纱。

染色：如果需要彩色毛条，它们必须在并条和纺纱前进行染色，迫使染液通过它们。进一步的针梳和精梳后，它们就可并条，纺纱了。

并条：并条工序的主要目的是经三道或四道工序逐渐降低毛条的细度，制成纺用粗纱。这是通过针梳机完成。粗纱机是梳理和细纱之间的中间工序，它将细条牵伸成适合纺细纱的细度，在纤维卷绕到大筒管前，既要增加几圈捻度，也要由罗拉轻搓条子。加捻或搓动作用给予细纤维集合体一定的抱合力以便于粗纱从筒管退绕下来，喂入细纱机。

细纱：这是最后一道加工工序，是通过牵伸来降低纤维条的细度。在精纺纱系统中，材料的牵伸倍数为 20，也就是说纤维集合体离开输出罗拉时比进入输送罗拉的长度要长 20 倍，细度细 20 倍。这是纤维所能经受的最高牵伸。针梳机的牵伸倍数通常是 8，粗纱机为 12。这也比粗梳纱高得多，粗梳纱的牵伸倍数通常小于 2，最终支数由梳理机确定。最终支数与所用纺纱牵伸有一定关系，与所用纤维类型也有关系，如合成纤维纺粗支数可牵伸到 35 倍以上。一旦纤维经过牵伸，纱线就通过钢丝圈被加捻和卷绕到筒管上。

Lesson 26　Texturing

Fabrics manufactured from filament silk, and indeed unmodified continuous filament synthetic yarns, have characteristics that are very different to yarns spun from staple fibres such as cotton or wool, staple synthetic fibres, or staple fibre blends such as cotton and polyester or wool and nylon. Continuous filament yarns, when simply drawn after spinning to produce desirable mechanical properties, tend to exhibit smoothness, evenness, and parallelism compared to the less regular, more bulky and hairy staple-fibre spun yarns.

Of course, continuous filaments from synthetic fibres can be cut into staple lengths through a separate filament cutting process to form filament staple suitable for a conventional spinning process using raw synthetic- or blended-staple fibres. However, this is a multi-stage process, which is costly, utilising an end-spinning process that was developed for natural fibres. As a consequence, in the 1950s and 1960s, processes referred to as "texturing" were developed. Texturing imparts desirable textile properties to continuous filament yarns without destroying their continuity through introducing distortions or crimp along their lengths.

Textured yarns have suitable volume, stretch and recovery, and air porosity for everyday use in fabrics for a wide range of textile end uses. Initially, the process comprised the insertion of twist into nylon or polyester yarns, thermally setting the twist by steaming in an autoclave, cooling and then untwisting the yarn by the same number of turns as originally inserted. Texturing processes are aimed at capitalising from raw material yarn properties rather than producing a material that fits the traditional yarn process routes. Thermoplastic, melt-spun continuous filament yarns provide the ideal platform for texturing, as imparted filament distortions can be softened through heat and set by cooling. The texturing process, therefore, evolved around key synthetic thermoplastic filament yarns, which soften when heated and reset when cooled, namely: polyester, nylon, polypropylene.

In the 1970s, various texturing methods were introduced commercially to the yarn manufacturing industry:

• Edge Crimping—drawing a thermoplastic yarn over a heated edge, creating differential internal stresses in the filament cross-sections. Edge crimp yarns exhibit high stretch; they were used for ladies hosiery and circular knit fabrics.

• Knit-de-Knit—knitting the thermoplastic yarn on a small diameter circular knitting machine. The plain knit fabric is then heat set and subsequently de-knitted and wound onto a package. After de-knitting, the yarn is deformed according to the knitted loop shapes, forming a three-dimensional structure.

• Stuffer-box—thermoplastic yarn is overfed into a heater cylinder, under a pressure from the feed-roller delivery that exceeds that of the controlled outlet resistance. As a result of this process, the filaments are heat-set in a buckled/crimped form before exiting the stuffer-box and being wound onto a package.

• False Twist Texturing—inserting high twist levels into thermoplastic yarns, setting the twist by heating and cooling prior to de-twisting.

• Air Jet Texturing—air-entangling of continuous filament yarns, applying overfeeds, draw and heat set(see Fig. 26.1).

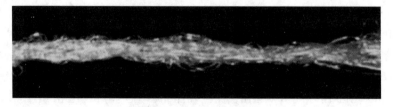

Figure 26.1 Air-textured yarn

1.Twist-texturing

The basis of twist-texturing is to twist a yarn as highly as possible, set it by heating and cooling, and then untwist it. This gives "a twist-free bundle of twist-lively filaments". In order to relieve the torque, the filaments snarl into "pigtails", which cause a large yarn contraction. The yarn can be stretched to over five times of its fully contracted length before the filaments are straightened out. Fabrics can be highly stretched, but come back when released. The filaments are in a helical configuration in the twisted yarn, and, after setting, they want to return to the crimped form. This dictates the initial form of fibre buckling. When a fully extended stretch yarn is allowed to contract by 10% to 20%, the filaments follow helical paths, which alternate from right-handed to left-handed. When a yarn is set in this form, it has high bulk and low stretch.

In the early days, spun LOY yarn tended to be pre-drawn on draw twisting machines after the spinning operation, to provide desirable, stable yarn mechanical properties for subsequent use in false twist texturing.

In the sequential method, the feeder yarn is pre-drawn within the process to attain the desired yarn tensile properties (tenacity and elongation). The older LOY materials in general had to be textured by this method because threading a low orientated/low crystalline yarn on the texturing heater is very difficult via the alternative simultaneous method. Pre-drawing provides both molecular orientation and crystallinity to support trouble-free threading via the sequential route. In simultaneous draw texturing, the feeder yarn is simultaneously drawn, twisted and heat-set prior to de-twisting in a continuous process. Since the introduction of POY spinning in the early 1970s, the simultaneous method has become the norm in false twist texturing. False twist texturing depends on setting fibres in one geometry and then changing to another, which generates stress that can be relieved by buckling.

2.Jet-screen texturing: BCF yarns

Another principle is to set fibres in the required crimped form. Another approach is adopted to produce BCF yarns, which are used as coarse carpet and upholstery yarns. Its origins lie in DuPont research on jets. Turbulent hot fluid produces an asymmetric shrinkage, which causes filaments to buckle.

生词与词组

1.blended-staple fibre 混合短纤维

2.multi-stage [ˈmʌltɪˌsteɪdʒ]*adj.* 多级的；多段的

3.distortion [disˈtɔːʃ(ə)n]*n.* 扭变；扭曲；扭转

4.textile end use 终端纺织品

5.autoclave [ˈɔːtəʊkleɪv]*n.* 高压锅；高压釜

6.edge crimp 刀口卷曲

7.de-knit [ˌdiːˈnɪt]*v.* 脱编；解编

8.stuffer-box 填塞箱

9.overfeed [ˈəʊvəfiːd]*v.* 超喂；过量

10.snarl [snɑːl]*v.* 缠结；纠缠

11.asymmetric [ˌeɪsɪˈmetrɪk]*adj.* 不对称的；非对称的

译文

第26课　变形纱

由长丝制造的织物，实际上是并未改变的连续长丝合股成纱线，其特性截然不同于由棉、羊毛、合成短纤维以及棉/聚酯、羊毛/尼龙等短纤维混纺纱纺成的纱线。与不规则、蓬松和多毛的短纤纱相比，纺丝后的连续长丝纱经简单拉伸获得所需的机械性能，往往表现出光滑、均匀和平行。

当然,合成纤维的连续长丝可通过长丝切割过程切割成短纤维长度,以形成适合合成短纤维或混合短纤维的常规纺纱工艺的长丝短纤维。然而,这是一个多流程的工艺,其成本高昂,使用的是为天然纤维开发的端纺工艺。因此,在20世纪50年和60年代,开发了被称为"变形纱"的工艺。变形赋予连续长丝纱线所需的纺织性能,而不会因沿其长度引入变形或卷曲而破坏其连续性。

变形纱具有合适的体积、拉伸和回复率以及孔隙率,适用于日常使用的各种纺织品。该过程先将尼龙或聚酯纱线加捻,再在高压釜中蒸汽热定型捻度,冷却,然后将纱线解捻。变形工艺旨在利用原材料纱线特性,而不是生产适合传统纱线工艺流程的材料。热塑性、熔纺连续长丝纱线为变形提供了理想的平台,其可通过加热软化使长丝扭曲再冷却固化。因此,变形工艺与合成热塑性长丝(聚酯、尼龙、聚丙烯)纱线紧密相关,加热时软化,冷却时复位。

20世纪70年代,各种变形方法被引入纱线制造行业:

刀口卷曲法——热塑性纱线通过加热的边缘,在长丝横截面中诱发不同的内部应力。卷边纱线具有高拉伸性;它们被用于制作女士袜子和圆机针织物。

假编法——在小直径针织圆机上编织热塑性纱线,将平针织物热定形后脱圈再绕成规定卷装。拆散的纱线具有针织线圈的形状,呈三维结构。

填箱法——进料辊输送压力超过受控出口阻力,热塑性纱线被超喂进入加热筒。该过程的结果是长丝在离开填塞箱、绕成卷装前以弯曲/卷曲形式进行热定型。

假捻变形法——热塑性纱线加高捻,以设定捻度通过加热、冷却,再解捻。

喷气变形法——超喂连续长丝,经空气缠结、拉伸和热定型(见图26.1)。

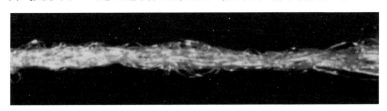

图 26.1　喷气变形纱

1. 加捻变形

加捻变形的基础是将纱线尽可能加高捻,加热定型,冷却,再将其解捻。这提供了"一束捻度不稳定的长丝"。为了降低扭矩,长丝缠结成"猪尾",导致纱线大幅收缩。在长丝被拉直之前,纱线可拉伸至其完全收缩长度的五倍以上。织物可大幅拉伸,释放后仍可恢复原状。在加捻纱线中长丝呈螺旋结构,定型后它们要恢复卷曲状态。这决定了纤维屈曲的初始状态。当完全伸展的弹力纱可收缩10%到20%时,长丝沿着螺旋路径移动,左右旋交替。当纱线以这种形式设置时,它具有高膨松度和低拉伸性。

早期,纺制的LOY纱线在纺丝后,经拉伸加捻机进行预拉伸,以提供理想、稳定的纱线机械性能,以便用于假捻变形。在顺序方法中,喂入纱被预拉伸以获得所需的纱线拉伸性能(强度和伸长率)。老的LOY材料一般必须以这种方法变形,因已被替代的同步法在变形加热器上难以穿入低取向/低结晶纱线。预拉伸提供分子取向和结晶度,纱线可按顺序法自由地穿入。在拉伸变形中,喂入纱进行拉伸、加捻和热定型连续工艺。20世纪70年代初,引入POY纺纱以来,顺序法已成为假捻变形的标准。假捻变形取决于纤维设置为一种几何形状,再转换为另一种几何形状,这会产生应力引起纤维屈曲。

2. 喷射丝网变形:BCF纱线

另一个原则是将纤维定型为所需的卷曲形式。采用另一种方法生产BCF纱线,其用于粗地毯和室内装潢用纱线。它源于杜邦的喷气法研究。湍急的热流体产生不对称收缩,导致长丝弯曲。

Lesson 27 Rope

Ropes are structures made of textile fibres. They are defined as approximately cylindrical textile bodies whose cross-sections are small compared to their lengths and they are used as tension members. The rope structure contains large numbers of synthetic or natural fibres in coherent, compact, and flexible configurations, usually to produce a selected breaking strength and extensibility with a minimum amount of fibres. For most common ropes, the fibres are arranged in helical structures whose axes form helices in larger structures, the process continuing in stages until a rope is complete. Either laying (twisting) or interlacing (braiding) techniques are used to arrange and contain the rope elements.

Unstructured ropes, or those with very small helix angles, may require enclosure techniques of braiding to contain the elements. Rope structures can be divided into two general categories. They are laid and braided ropes with fairly high twist or braid angles; these are the most common structures for general purpose use. They are found in industrial, marine, recreation and general utility service. They cover everything from small cords used to tie-up packages, to large hawsers for mooring tankers, and include clothes-lines, yachting ropes, haul lines for fishing nets, lifting slings, and a host of other applications. Laid and braided ropes (see Fig. 27.1) for general purpose uses are:

- Three-strand, four-strand and six-strand laid;
- Eight-strand plaited;
- Double braid (also called "braid-on-braid" and "2-in-1 braid");
- Solid braid (also called "parallel braid").

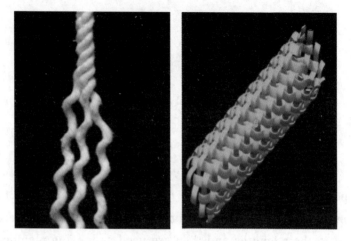

Figure 27.1　The structures of laid and braided ropes

Low twist rope structures are used for specialised and demanding applications where high strength to weight ratios and low extensibility are essential. This would include tethers for astronauts, guys for tall masts, deep sea salvage recovery ropes, mooring lines for floating oil platforms, and hoist cables for deep mines. Low twist ropes for specialised uses are:

- Braided rope with jacket;
- Parallel strand, jacketed;
- Parallel yarn (or filament), jacketed;
- Wire rope type, exterior jacketed;
- Wire rope type, each strand jacketed.

There are wide variations within each type of rope structure. The appearance and behaviour of the structure are influenced by many factors, some of which are:

- Type and size of fibre material (textile yarns) used;
- Dual or blended fibre materials;
- Amount of twist at each stage of production;
- Size and number of subcomponents at each stage of production;
- Tension on subcomponents during production;
- Post-production heat and other treatments.

Individual companies have evolved their own technology and rope designs into more of a skill than a science. Personal preference and company history are subjective factors that infiuence the choice of rope structure, but an important objective factor is the availability of production machinery. Ropes get tweaked this way and that, but in the end, for good performing general-purpose ropes, the range of design variations within any one type of construction is fairly small.

生词与词组

1.interlace [ˌɪntəˈleɪs]*v.* 交织；交错

2.braid [breɪd]*v.* 编织；编结

3.marine [məˈriːn]*adj.* 海运的；海事的

4.tie-up [ˈtaɪˌʌp]*v.* 捆绑；打结

5.hawser [ˈhɔːzə(r)]*n.* 缆索

6.mooring tanker 泊油轮

7.yachting [ˈjɒtɪŋ]*n.* 游艇

8.haul [hɔːl]*v.* 拖；拉

9.sling [slɪŋ]*n.* 吊索；悬带

10.strength to weight ratio 比强度

11.extensibility [ɪkˌstensəˈbɪlətɪ]*n.* 可伸展性；可延长性

12.astronaut [ˈæstrəˌnɔt]*n.* 宇航员

13.mast [mɑːst]*n.* 船桅

14.salvage [ˈsælvɪdʒ]*n.* 抢救，救援

15.hoist [hɔɪst]*v.* 升起；吊起

16. subcomponent [ˌsʌbkəmˈpəʊnənt]*n.* 子部件；亚成分

17.tweak [twiːk]*v.* 拧；扭；稍稍调整

译文

第 27 课　绳索

绳索是由纺织纤维制成的。它们定义为近似圆柱形的纺织体,其横截面远小于其长度,它们用于受拉构件。绳索结构包含了大量的合成纤维或天然纤维形成连贯、紧密、易弯形态,通常用最少量纤维制造预定断裂强度和延展性的绳索。多数绳索的纤维排列呈螺旋结构,其轴在更大的结构中形成螺旋,该过程分阶段继续,直到完成绳索。使用搓捻(扭曲)或交织(编织)技术来排列和容纳绳索单元。

非结构化绳索,或那些小螺旋角的绳索,可能需要环绕编织技术来容纳单元。绳索结构可分为两大类。它们是具有高扭曲的搓绳和高编织角度的编绳,这些是常规用途的最常见结构。它们涵盖了从捆绑包裹用的绳到用于系泊油轮的大型缆索,包括晾衣绳、游艇绳索、渔网牵引绳、起重吊索和其他应用。一般用途的搓捻和编织绳索(如图 27.1)是:

（1）三股、四股、六股搓合;

（2）八股编织;

（3）双编织(也称为"编织对编织"和"二合一"编织);

（4）实心编织(也称为"平行编织")。

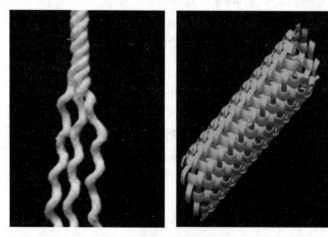

图 27.1　搓绳和编绳的结构

低捻度绳索结构,用于高比强度和低延展性的专业和要求高的场合。这包括宇航员的系绳、高桅杆的拉索、深海打捞回收绳、漂浮石油平台的系泊绳以及深井的起重钢缆。专门用途的低捻绳有：

（1）包芯的编织绳；

（2）包缠平行股；

（3）包缠平行纱（丝）；

（4）钢丝绳型，外包缠；

（5）钢丝绳型，每股包缠。

每种类型的绳索结构都有很大的差异，其结构的外观和行为受到诸多因素的影响，其中一些因素是：

（1）所用纤维材料（纱线）的类型和尺寸；

（2）双纤维或混合纤维材料；

（3）每阶段的捻度；

（4）每个生产阶段的子部件的尺寸和数量；

（5）生产环节子部件的张力；

（6）后期热处理和其他处理。

个别公司已将它们自己的技术和绳索设计发展成为一种技能，而不是一门科学。个人喜好和公司历史是影响绳索结构选择的主观因素，但一个重要的客观因素是可用的生产机械。绳索会以这样或那样的方式进行调整，但最终，对于性能良好的通用绳索，任何一类结构设计的变化范围都相当小。

Lesson 28　Yarn testing standards and parameters

Textile testing forms an essential link in the quality assurance and quality control chain and may be divided broadly into two major types, namely subjective and objective. Clearly, the intended application of the yarn or fabric is critical in determining which properties, and therefore tests, are important. For example, for fabric to be used in children's nightwear, flammability (e.g., the Limiting Oxygen Index—LOI) would be of paramount importance, whereas for fabrics to be used in parachutes, bursting and tear strength, impact resistance and air permeability would be critically important.

Nevertheless, generic or general testing of yarns and fabrics is often carried out irrespective of their end use so as to ensure consistency in the yarn and fabric and to detect any changes suggesting possible production problems. Ultimately, wearer (field) trials would provide the most reliable measure of actual performance, but these are expensive, time consuming and difficult to organise and design in such a way that meaningful and accurate results are obtained.

In carrying out any test it is important to as far as possible use accepted test methods ensuring that the correct sampling, sample preparation and handling, atmospheric conditions (e.g., 20℃ ±2℃ and 65% ± 2% RH), conditioning time and instrument calibration are adhered to. Standard test method organisations include:

· AATCC (American Association of Textile Chemists and Colourists)

· ASTM (American Society for Testing of Materials)

· BSI (British Standards Institution)

· BIS (Bureau of Indian Standards)

· DIN (Deutsche Industrie Norm, German National Standards)

· EN (European Norm)

· ISO (International Standards Organization)

· JIS (Japanese Industrial Standards)

It should be noted that BSI, EN and DIN standards now largely use ISO designations.

1.Yarn testing

The yarn represents the final outcome of the fibre processing or yarn manufacturing part of the fibre to fabric textile pipeline. Prior to the yarn it is still possible, within certain limits, to take corrective action should problems or mistakes have occurred during the earlier processing stages. Once the yarn has formed, any corrective action is limited to yarn clearing (i.e., removal of gross or unwanted faults), uptwisting and singeing. The properties of the yarn largely determine subsequent fabric manufacturing performance and efficiency (also sewing performance) as well as the properties, aesthetic and functional, of the fabric and end product. Yarn quality control, and the associated testing of those yarn properties which determine the yarn quality, are therefore crucial in ensuring that the yarn meets the requirements of the subsequent fabric manufacturing stages as well as of the fabric and end product.

2.Count (linear density)

Count (linear density) is a fundamental structural parameter of a yarn, and the average count and its variations are of paramount importance in virtually all facets of textile performance and specifications. Undue variations in yarn count, or off-specification count lead to problems in terms of fabric mass and barré, and often to claims.

Traditionally, the average yarn count (linear density) and its variation (CV) are determined by weighing a length of 100m of yarn. Linear density (count) variations (CV), based upon the weighing of 100m yarn, of less than 2% are generally considered acceptable. Ideally, 20 × 100m of yarn, preferably each coming from a different yarn bobbin or package, should be tested. The count of cotton yarn can be expressed in any one of the following three systems:

1. Tex count (linear density): weight in grams of 1,000m of yarn (i.e.,mg/m or g/km);

2. Cotton count (Ne): number of 840yd in a pound;

3. Metric count (Nm): number of 1,000m of yarn in one kilogram.

3.Twist

Twist is one of the fundamental constructional parameters for yarns. Average twist levels and twist variations are reflected in the yarn tensile properties, thickness, bulk, stiffness and handle. These, in turn, are reflected in the yarn performance during the subsequent fabric manufacturing processes and ultimately in the corresponding fabric properties. Traditionally, twist was counted manually by the direct twist counting method where the operator would clamp the yarn test length (popularly 25mm for single yarn and 500mm for two-ply yarn) under a pre-determined tension, and then untwist the yarn, using a "dissecting needle" until all the twist is removed. Today, twist is more commonly measured automatically, using different test methods and techniques, for example, double-untwist-retwist tests, untwist-retwist tests described in ASTM D-1422, multiple untwist-retwist and twist-to-break, the test length popularly being 50cm. Typically 100 tests per yarn lot are measured, it also being possible to measure the twist (helix angle optically). The yarn twist is expressed in terms of turns per metre (or turns per centimetre) and the CV of twist in percentage, noting the gauge length. Twist results may be assessed using "norms or standards", such as the Testex Twist Statistics, variation in yarn twist being particularly important. Various instruments and test methods are available for testing yarn twist.

4.Evenness

The terms yarn evenness, unevenness, regularity and irregularity are often used interchangeably and taken to refer to the same yarn characteristic, namely the uniformity or evenness of the yarn mass per unit length (linear density), measured capacitively, or of the yarn cross-sectional size or diameter, measured optically. This characteristic is important from two perspectives, namely from an aesthetic or appearance point of view and from a technical or functional (performance) point of view, since unevenness (i.e., thinner or thicker segments) in the yarn cross-section, is commonly associated with variations in the yarn strength and elongation. In practice, yarn evenness is most frequently measured using capacitance, as opposed to optical, techniques which means that it is assessed in terms of changes in segment mass (linear density) rather than in terms of cross-sectional volume (bulk) or diameter, assuming the dielectric properties (notably moisture and fibre blend) of the yarn are constant.

5.Hairiness

Hairiness normally refers to fibre ends, fibre loops and fibres of varying length (wild fibres) protruding from the surface (or body) of the yarn, frequently as fibre tails. Although hairiness is desirable in certain types of fabrics, notably soft knitteds, brushed fabrics and flannels, it is undesirable in other fabrics, such as shirting. Yarn hairiness affects the fabric surface appearance and properties, including fabric pilling, handle and comfort (thermal insulation). Yarn hairiness also plays an important role in weavability, since the protruding hairs tend to catch on adjacent yarns causing yarn breakages and loom stoppages. It is particularly important in air jet weaving. In fact, one of the main objectives of sizing is to smooth down the hairs on the yarn, i.e., to reduce yarn hairiness, thereby improving yarn weaving performance. Yarn hairiness positively affects the heat and wear generated when the yarn runs over metal or other surfaces, such as travellers and yarn guides. Variations in yarn hairiness can be particularly problematic, being reflected in variations in fabric surface appearance, and can cause fabric barre and streakiness, it therefore being important that yarn hairiness remains acceptably constant within a certain application and yarn consignment. It is now possible to measure yarn hairiness on-line, for example, on programmable yarn clearers, Yarn Hairiness Grades helping to visualise various levels of hairiness.

生词与词组

1.nightwear [ˈnaɪtweə(r)]*n.* 睡衣；广告衫

2.flammability [ˌflæməˈbɪlɪti]*n.* 可燃性

3.Limiting Oxygen Index 极限氧指数

4.parachute [ˈpærəʃuːt]*n.* 降落伞

5.air permeability 透气性

6.atmospheric [ˌætməsˈferɪk]*adj.* 大气的；大气层的

7.calibration [kælɪˈbreɪʃ(ə)n]*n.* 标定；校准

8.pipeline [ˈpaɪplaɪn]*n.* 管道

9.gross [grəʊs]*n.* 全部 *adj.* 总的

10.uptwisting [ˈʌpˌtwɪstɪŋ]*n.* 上行式捻线；上捻

11.singeing [ˈsɪndʒɪŋ]*n.* 烧毛；燎毛

12.clamp [klæmp]*n.* 夹具；夹子

13.dissecting needle 分析针；解剖针

14.double-untwist-retwist 二次退捻加捻法

15.untwist-retwist 退捻加捻法

16.twist-to-break 扭断

17.dielectric [ˌdaɪɪˈlektrɪk]*n.* 电介质 *adj.* 介电的

18.loop [luːp]*n.* 环形；环状物；圈

19.flannel [ˈflænl]*n.* 法兰绒

20.loom stoppage 织机停机

21.streakiness [ˈstriːkɪnɪs]*n.* 条花；竖条纹

22.consignment [kənˈsaɪnmənt]*n.* 递送；托运

23.programmable [ˈprəʊɡræməbl]*adj.* 程控的；可编程序的

译文

第28课　纱线测试标准及参数

纺织品检测是质量保证和质量控制链中必不可少的一个环节，大致可分为主观、客观两大类。显然，纱线或织物的预期应用关键取决于性能，因此测试是重要的。例如对于用于儿童睡衣的织物，可燃性（如极限氧指数，LOI）是至关重要的，而对于用于降落伞的织物，顶破和撕裂强度、抗冲击性和透气性则是至关重要的。纱线和织物的通用或一般测试往往与其最终用途无关，为确保纱线和织物的一致性，检测任何因生产问题引起的变化。最后，试穿（现场）提供了最可靠的实际性能测量，但以此方法获得有意义、准确的结果，往往昂贵，耗时且难以组织和设计。

进行任何测试时，尽可能使用公认的测试方法很重要，以确保取样、样品制备和处理、大气条件（如温度20℃±2℃，湿度65%±2%）、调节时间和仪器校准等准确。标准测试方法组织包括：美国纺织化学和颜色协会、美国材料测试协会、英国标准协会、印度标准局、德国工业规范（德国国家标准）、欧洲标准、国际标准化组织、日本工业标准化组织。应该注意的是，英国标准协会、欧洲标准和德国工业规范标准现在大部分使用国际标准化组织标识。

1.纱线测试

纱线代表了纤维加工或纤维到纤维加工纱线至织物流水线的最终结果。在成纱之前，在一定限度内仍然可采取措施，纠正早期加工阶段出现的问题或错误。一旦纱线形成，任何纠正措施仅限于清纱（即去除粗大或不需要的缺陷）、上捻和烧毛。纱线性能在很大程度上决定了后续织物的织造性能和效率（也包括缝纫性能），以及织物和最终产品的性能、美感和功能。因此，纱线的质量控制及纱线质量的关键性能的相关测试，对于确保纱线满足后续织物生产阶段以及织物和最终产品的要求至关重要。

2. 支数（线密度）

支数（线密度）是纱线的基本结构参数，平均支数及其变化在纺织品性能和规格的几乎所有方面都至关重要，纱线支数的过度变化或不合格的支数会导致织物质量及索赔问题。

我们习惯通过称量 100 米长的纱线重量来确定平均纱支数（线密度）及其变化（CV）；100 米长纱线其线密度变化（CV）小于 2% 一般来说是可以接受的。理想情况下，应该测试 20 个 ×100 米长度的纱线，且纱线最好来自不同的管纱或筒纱。棉纱支数可以用以下三种体系中的任意一种来表示：

（1）特克斯支数（线密度）：1000 米长纱线的重量克数（即毫克 / 米或克 / 千米）；

（2）英制支数（Ne）：840 码长纱线的重量磅数；

（3）公制支数（Nm）：1000 克纱线含有 1000 米长度纱线的个数（1克重纱线所具有的长度米数）。

3. 捻度

捻度是纱线的一个基本结构参数。平均捻度和捻度不匀率反映了纱线的拉伸性能、粗细、膨松度、刚度和手感。这又反映了后续织物制造过程中的纱线性能以及最终相应的织物特性。传统上，捻度是由直接捻度计数法手动计算的，操作员将预定的张力下测试长度的纱线（单股纱线通常为 25 毫米，两股纱线通常为 500 毫米）夹紧，然后解捻纱线，使用"分析针"直到所有捻度退掉。现在，捻度常用不同的测试方法和技术自动测量，如 ASTM D-1422 中描述的二次解捻加捻测试、解捻—加捻测试、多次解捻—加捻和捻断，测试长度通常为 50 厘米，通常每批纱线进行 100 次测试，也可测量捻度（光学螺旋角）。纱线捻度以每米的捻回数（或每厘米的捻回数）表示，捻度不匀率以百分数表示（注意标尺）。捻度结果可以使用"规范或标准"进行评估，如 Testex 捻度统计，纱线捻度的不匀率尤为重要。多种仪器和测试方法可用于测试纱线捻度。

4.条干均匀度

纱线的均匀度、不均匀度,规则性和不规则性这些术语往往可互换,因它们都指纱线的同一个特性,即电容测试单位长度重量(线密度)纱线的一致性或均匀性,或者光学测量纱线横截面的尺寸或直径。从两个角度来看,这一特性很重要,即从美学或外观的角度和从技术或功能(性能)的角度来看,因为纱线横截面的不均匀(即较细或较粗的部分)通常与纱线强度和伸长率的不匀率相关。实际中,纱线均匀度最常用电容测量,而不是光学测试,假设纱线的介电特性(特别是水分和纤维混合物)是恒定的,这意味着它是依据片段质量(线密度)的变化而不是根据横截面体积(体积)或直径的变化来评估的。

5.毛羽

毛羽通常是指从纱线表面(或主干)突出的不同长度的纤维末端、纤维环和纤维,通常是纤维尾部。尽管在某些类型织物中,特别是柔软的针织织物、拉丝织物和绒织物,毛羽很受欢迎,但在其他织物如衬衫中,毛羽是不受欢迎的。纱线毛羽会影响织物表面外观、性能,包括织物起球、手感和舒适性(隔热)。纱线的毛羽对织造也有重要影响,突出的毛羽易与相邻的纱线粘连,导致纱线断头和织机停机。在喷气织机中尤为严重。事实上,浆纱的主要目的之一是贴服纱线上的毛羽,即减少纱线毛羽,从而提高纱线的织造性能。当纱线在金属或其他表面(例如钢丝圈和导纱器)上运行时,纱线毛羽易生热,磨损。纱线毛羽的变化特别成问题,可反映在织物表面外观变化上,也可能引起织物横档和条纹,因此在特定应用和纱线托运中,纱线毛羽保持恒定、可接受是很重要的。现在,纱线毛羽已可在线测量,如可编程清纱器,毛羽分级系统可视觉识别毛羽等级。

Lesson 29 The development of weaving technology

Fabric is generally defined as an assembly of fibres, yarns or combinations of these. Fabrics are most commonly woven or knitted, but the term includes assemblies produced by lace making, tufting, felting, and knot making as well as by the so-called non-woven processes. The raw materials used and the machinery employed mainly govern the type of fabrics produced. Of all the fabric formation procedures, weaving is the oldest method and it is probably as old as human civilization. Historical evidences show that Egyptians made woven fabrics some 6,000 years ago and Chinese made fine fabrics from silk over 4,000 years ago. Woven fabric is produced by interlacing of threads placed perpendicular to each other. The yarns that are placed length-wise or parallel to the selvedges (edges) of the cloth are called warp yarns. The yarns that run cross wise are called weft or filling yarns. There are numerous ways of interlacing yarns to produce a variety of fabric structures.

In different civilizations, it is believed, various types of handlooms were invented for weaving cloth. Weaving was a cottage industry until John Kay invented the hand-operated fly shuttle in the year 1733. Edmund invented the power loom in 1785. Introduction of the electric power driven loom opened up a new vista and particular emphasis was placed on increasing the loom speeds and thereby achieving higher productivity. Besides, developments in science have significantly changed the technologies of inserting weft yarns whereby not only productivity increased but also newer weaving processes and product developments took place. As a result of these advances, shuttleless and multiphase looms emerged in the 20th century. The various weaving processes are dealt with in the same sequence as they occur in the weaving of a fabric.

生词与词组

1. knot making 打结；编结
2. selvedge [ˈselvɪdʒ]*n.* 织边；镶边
3. cross wise 横向
4. weft [weft]*n.* 纬线；纬纱
5. filling yarn 纬纱
6. interlacing [ˌɪntə(:)ˈleɪsɪŋ]*n.* 交错；交织
7. handloom [ˈhændˌluːm]*n.* 手工织布机；手织机
8. cottage [ˈkɒtɪdʒ]*n.* 小屋；村舍
9. shuttle [ˈʃʌtl]*n.* 梭子 *v.* 穿梭；穿梭移动
10. multiphase loom 多梭口织机；有梭织机

译文

第 29 课　机织技术发展

织物通常被定义为纤维、纱线或这些的组合。织物一般为机织或针织，但该术语也包括以花边、簇绒、毡和打结以及非织造工艺生产的集合体。所用的原材料和机器主要决定了所生产织物的类型。在所有织物形成过程中，织造是最古老的方法，它可能与人类文明一样古老。历史证据表明，6000 年前埃及人制造了机织织物，4000 多年中国人用丝绸制造了精美的织物。机织物是由相互垂直的线交织而成的。沿长度方向或平行于布料的织边（边缘）的纱线称为经纱。横向运行的纱线称为纬纱。多种交织纱线的方法可生产各种织物结构。

人们相信，在不同的文明中，人们发明了各种类型的手摇织机来织布。在凯伊发明手动飞梭（1733 年）之前，织造是手工业。1785 年，埃德蒙发明了动力织机。电动织机的引入开辟了新的前景，特别强调提高织机速度,从而实现更高的产能。此外,科学的发展极大地改变了纬纱插入技术,不仅提高了生产率,而且还更新了织造工艺和产品开发。由于这些进步,20世纪出现了无梭织机和多相织机。各种织布过程的处理顺序基本一致。

Lesson 30 Warping, warp sizing, drawing-in and denting

1.Warping

The objective of preparing a warp is to supply a sheet of yarns, of desired length to the succeeding processes, laid parallel to each other. Warping is done by winding a number of yarns from a creel of single-end packages such as cones or cheeses onto a beam. The warp beam that is installed on a weaving machine is called a weaver's beam. The two basic systems of preparing warp are known as the direct system and the indirect system. Direct warping is used in two ways. First, it can be used to produce directly the weaver's beam in case of strong yarns and when the number of warp ends on the warp beam is relatively small. Second, it can be used to produce smaller intermediate beams, and the warper's beams are combined later at sizing stage to produce the weaver's beam. Indirect warping is used to produce a section beam which facilitates a wide range of fabric constructions that would require the use of multi-coloured yarns, fancy weaves and assortment of yarn counts. Warp yarn is wound onto the beam in sections, beginning with the tapered end of the beam. Because of the geometry of the concentrically wound yarn sections, the end of the beam is tapered to ensure the yarn on the beam is stable.

Today's warping machines can process all kinds of materials including coarse and fine filament and staple yarns, monofilaments, textured as well as silk.

2.Warp sizing

Sizing gives a protective coating to warp yarns for them to withstand the tension and abrasion that yarns undergo during the weaving process. In other words, sizing increases the strength and reduces hairiness of yarn and at the same time helps maintain the required flexibility and elasticity. Thus, sizing is done to attain a high weaving efficiency by reducing warp breaks during weaving. Sizing is the operation of coating of a polymeric film forming agent (called size) on the warp yarns. Generally, the size mix contains film forming agents (e.g., starch, PVA), lubricants like mineral oils, paraffin wax, humectants such as ethylene glycol, glycerol, etc., preservatives, water and defoamer.

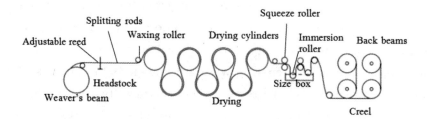

Figure 30.1　Multi-cylinder sizing machine

The major parts of sizing machines are the creel, size box, drying units, beaming and various control devices (Fig. 30.1). After leaving the warper, the beams are placed on the creel and the sheet of yarn from the creel is passed through the size box which contains the sizing solution. The yarns pick up the required quantity of size solution and pass through the squeezing rolls where any excess size is squeezed off. In the next process when the yarns pass through drying section, most of the water from the warp evaporates and the yarns are wound on the weaver's beam. Then, the weaver's beam goes through drawing-in and denting. Once the fabric is woven, the size is removed by desizing process, except in a few cases where it is loom finished material.

3.Drawing-in and denting

Drawing-in and denting form an essential link between the designing of a fabric and the working parts of the loom. Drawing-in is the threading of warp yarns from the sized weaver's beam into the eyes of the healds which are mounted in the heald frames in the loom. The threading of yarns follows the desired pattern of the fabric. Each warp yarn end will be drawn through the eye of one heald. Healds which are required to be raised at the same time will be placed on the same heald frame, while healds that lift differently will be placed on different heald frames. Healds control the movement of warp threads to separate themselves into two layers so as to make a tunnel for filling or weft yarns to be inserted in the gap. This opening is called the shed. After the warp yarns in the beam are exhausted and if there is no change in the design, the corresponding ends of yarns of old and new warp beams are tied together and this is called tying-in process.

Denting is the arrangement of warp ends in the reed space. The reed is made of flat metal strips fixed at uniform intervals on a frame to form a closed comb-like structure. The spaces between the metal strips are known as dents. Reeds are identified by a reed number which is the number of dents per unit width. The reed holds one or more warp yarns in each dent. Denting plans describe the arrangement of the warp ends in the reed which controls the warp yarn density in the fabric. Warp density is expressed as either ends per inch or ends per centimetre. The main functions of the reed are to hold the warp yarns at uniform intervals, beat up the newly inserted weft and simultaneously support the shuttle during its traverse motion.

生词与词组

1.warp [wɔːp]*n.* 经纱
2.creel [kriːl]*n.* 粗纱架；经轴架；筒子架

3.single-end [ˈsɪŋglˈend]*adj.* 单头的；单端的

4.beam [bi:m]*n.* 经轴；织轴

5.intermediate [ˌɪntəˈmi:diət]*adj.* 中间的 *n.* 中间体

6.multi-coloured yarn 多色纱线

7.assortment [əˈsɔ:tmənt]*n.* 分类；各种

8.tapered end 锥形端

9.concentrically [kənˈsentrɪklɪ]*adv.* 同心地；同轴地

10.size mix 浆料混合物

11.paraffin wax 石蜡

12.humectant [hju:ˈmektənt]*n.* 湿润剂；保湿剂

13.ethylene glycol 乙二醇

14.glycerol [ˈglɪsəˌrəʊl]*n.* 甘油

15.preservative [prɪˈzɜ:vətɪv]*n.* 防腐剂；保存剂

16.defoamer [di:ˈfoʊmə(r)]*n.* 消泡剂

17.squeezing roll 挤压辊

18.squeezed off 挤掉

19.drying cylinder 烘筒

20.waxing roller 上蜡辊

21.splitting rod 分绞辊

22.headstock 车厢头

23.adjustable reed 伸缩筘

24.size box 上浆槽

25.immersion roller 浸压辊

26.evaporate [ɪˈvæpəreɪt]*v.* 使蒸发；使脱水

27.denting [ˈdentɪŋ]*n.* 穿筘

28.desizing [dɪˈsaɪzɪŋ]*n.* 退浆；退浆工艺

29.heald [ˈhi:ld]*n.* 综丝

30.frame [freɪm]*n.* 综框；画面（框架）

31.tunnel [ˈtʌnl]*n.* 通道；隧道

32.shed [ʃed]*n.* 梭口

33.tying-in process 打结过程

34.reed [ri:d]*n.* 筘；钢筘

35.strip [ˈstrɪp]*n.* 条；片

译文

第30课   整经、上浆、穿结经

1. 整经

准备经纱的目的是将所需长度的片状平行排列的纱供给后续工序。整经是将多根纱线从筒子架的各单头卷装(如圆锥形筒子或圆柱形筒子)上退绕,并卷绕到经轴上。安装在织机上的经轴称为织机经轴。经纱准备分为直接系统和间接系统两种。直接整经有两种方法。首先,当纱线强度高且经轴上的经纱数量较少时,它可直接用作织布机经轴。其次,它可用于生产较小的中间经轴,经轴的经纱在上浆阶段合并后形成织布经轴。间接整经用于生产分条经轴,其利于生产多色纱线、花式编织和各种纱线支数类型的织物结构。经纱从经轴的锥形端开始,分条绕在经轴上。由于同心缠绕纱线片段,经轴末端呈锥形以确保经轴上的纱线稳定。

现在的整经机可加工各种材料,包括粗细长丝和短纤纱、单丝、变形丝以及丝绸。

2. 经纱上浆

上浆为经纱提供了一层保护涂层,使它们能承受织造过程中的张力和磨损。换句话说,浆纱增加了纱线的强度,减少了毛羽,同时也有助于保持其所需的柔韧性和弹性。因此,上浆可提高织造效率,减少织造环节的经纱断头。上浆是在经纱上涂覆聚合物成膜剂(称为"上浆")的操作。通常,浆料混合物包含成膜剂(如淀粉、PVA)、润滑剂(如矿物油、石蜡)、保湿剂(如乙二醇、甘油等)、防腐剂、水和消泡剂。

浆纱机的主要部件是筒子架、浆纱箱、干燥装置、经轴和各种控制装置(图30.1)。离开整经机后,经轴放置在经轴架上,经轴架上的纱片通过装有浆液的浆纱箱。纱线沾取所需的上浆液并通过挤压辊,挤出多余的浆液。在接下来的工序中,纱线通过干燥区,经纱中的水分大部分蒸发,再缠绕在织机的经轴上。然后,织机的经轴再穿综和穿筘。织物一旦织好,再通过退浆工艺退掉浆液,除了在织物是织机成品材料的少数

情况下。

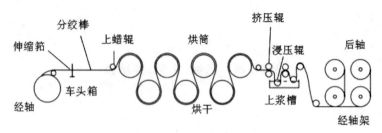

图 30.1　多滚筒浆纱机

3. 穿综、穿筘

　　穿综、穿筘是织物设计与织机工作部件之间的重要纽带。穿综是将经纱从浆纱机的经轴穿入安装在织机综框上的综丝眼。穿纱应遵循织物的图案要求。每根经纱都应穿入一个综眼。需要同步升起的综丝放置在同一综框上，不同步升起的综丝放置在不同的综框上。综丝控制经线的运动，将经纱分成两层，从而形成一个通道，用于纬纱插入。这个开口叫作梭口。经轴中的经纱用完后，如设计没有变化，则将新旧经轴纱线头的端系在一起，这称为打结过程。

　　穿筘是将经纱排列在筘隙间。筘片由扁平金属条以均匀间隔固定在框架上，形成封闭的梳状结构。金属条之间的空隙称为筘齿。筘片按筘片号进行标识，筘片号是单位长度的筘齿数量。筘的每个筘齿中可穿入一根或多根经纱。穿筘设计表示了经纱在筘中的排列规律，它控制着织物中的经纱密度。经纱密度表示为每英寸线数或每厘米线数。筘的主要功能保持经纱固定的间隔，击打新插入的纬纱，同时支撑横向运动的梭子。

Lesson 31　Weaving process

1.Weaving operation

The loom beam stores warp yarns. It is placed at the back of the loom. The yarns from the beam pass round the back rest roller which ensures that the yarns are maintained at the same weaving angle as the weaver's beam decreases in size during weaving. The warp let-off mechanism unwinds the warp yarns from the beam, as the yarns are woven into fabric, at the desired rate and at constant tension as required. The schematic diagram (Fig. 31.1) illustrates how the warp yarns pass through a loom. The yarns from the back rest roller are brought forward. Each end of the yarns from the beam is threaded through the eye of a drop wire.

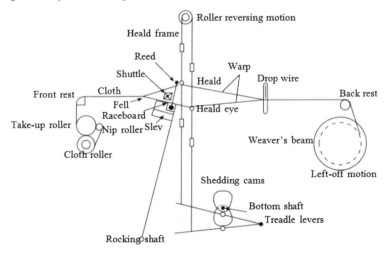

Figure 31.1　A cross-section through the loom

The drop wire stops the loom when a break occurs in any of the warp yarns. From here, yarns pass first through the eye of the heald and then through the dents of the reed. The operation that raises and lowers the heald frames according to the fabric pattern is known as shedding. In front of the reed, a triangular warp shed is formed by the two warp sheets and the reed. After the shed has been formed, the shuttle carrying the weft yarn traverses across the fabric. This is known as picking (weft insertion). One single strand of weft is known as a pick.

2.Weaving mechanism

To produce woven fabric, a loom requires three primary motions, shedding, picking and beating-up. Apart from these, there are two secondary motions in weaving, let-off and take-up motions. In an ordinary power loom all motions are operated by the main shaft called the crankshaft. The crankshaft is driven by an electric motor. One revolution of the crankshaft operates various functions of the loom at different time intervals.

3.Cam shedding

The shedding mechanism is operated by two shedding cams. The shedding cams are mounted on the bottom shaft in the case of plain weave design which needs to employ only two heald frames. However, wherever the design requires operating more than two heald frames, separate tappet shafts fitted to the bottom shaft are used. The shedding cams are mounted on the tappet shafts.

4.Dobby shedding

Dobby mechanisms can control up to 30 heald frames. There are two types of dobby mechanisms, the negative dobby and the positive dobby. In negative dobby shedding, the dobby lifts the heald frames which are lowered by a spring motion. In positive dobby shedding, the dobby raises and lowers the heald frames and the springs are eliminated. Negative or positive dobbies are further classified as single lift and double lift. The double lift dobby's cycle occupies two picks and therefore most of its motions occur at half the loom speed which allows higher running speeds. All modern negative dobbies are double lift dobbies, although this type of dobby has been largely replaced by the positive dobby.

5.Jacquard shedding

The Jacquard machine originally invented by Joseph served as the prototype for a very wide range of weaving and knitting machines in the textile industry. When a big or complicated design is to be made in weaving, jacquard shedding is used. In this shedding the warp ends are controlled individually by harness cords and there are no heald frames. There will be as many cords as there are ends in the warp, which enables unlimited patterns to be woven.

6.Picking

In shuttle weaving, the weft yarn is inserted by a shuttle which continuously moves back and forth across the width of the loom. A picking stick on each side of the loom activates the shuttle by hitting it and making it fly across the loom inside the open shed. Picking cams are mounted on the bottom shaft of the loom and are set at 180° to one another. When the picking cam rotates, it displaces the picking shaft via a cone. This makes the picking shaft pull the picking stick with a lug strap and by doing so the picking stick accelerates and hits the shuttle. Different mechanisms for picking operations are developed to suit the type of looms.

7.Let-off motion

As the yarn is woven, a let-off mechanism releases the warp yarn from the weaver's beam and at the same time maintains an optimum tension by controlling the rate of flow of warp yarns. If the tension of the warp yarn is not controlled at the desired level, warp breakage rates increase, which will affect the dimensional and physical properties of the fabric.

8.Take-up motion

Once the reed recedes after beating up the weft, the woven cloth is removed from the weaving area by the take-up motion. The take-up roller removes the cloth at a rate that controls weft density and the woven cloth is wound onto a cloth roller. The take-up roller is covered with perforated steel fillet or hard rubber depending upon the type of fabric being woven. The drive to the take-up roller is by a series of gear wheels which control the pick density. Presently, many modern loom manufacturers use electronic take-up motion which gives better and more accurate control of the pick spacing by means of a servo motor.

生词与词组

1.let-off ['letɒf]*adj.* 绕过的；松开的；送经的
2.triangular [traɪˈæŋgjələ(r)]*adj.* 三角形的
3.beating-up ['biːtɪŋˈʌp]*n.* 打纬
4.take-up ['teɪkʌp]*adj.* 卷曲的；拉紧的
5.shaft [ʃæft]*n.* 轴；柄；杆
6.crankshaft ['kræŋkʃæft]*n.* 曲轴；机轴
7.shedding ['ʃedɪŋ]*n.* 开口，梭口
8.cam [kæm]*n.* 凸轮
9.heald frame 综框
10.tappet shaft 挺杆轴

11.dobby [ˈdɒbɪ]*n.* 多臂机；小提花织物

12.Jacquard machine 提花机

13.harness cord 综线；通丝

14.reed recede 钢筘退出

15.servo [ˈsɜ:vəʊ]*n.* 伺服系统；伺服电机

16.nip roller 压辊

17.rocking shaft 摇臂轴

18.drop wire 停经片

19.heald eye 综眼

译文

第 31 课　机织过程

1.织造操作

　　织机经轴储存经纱，它放置在织机的后面。来自经轴的纱线绕过后梁辊，以确保其在织造过程中织造角度不随经轴卷装纱线尺寸的减小而变化。当纱线以所需的速率和恒定的张力织造成织物时，送经机构从经轴上退绕经纱。图 31.1 说明了经纱如何穿过织机。后梁辊的纱线引导向前。经轴的经纱都穿过停经片（吊丝）的孔。

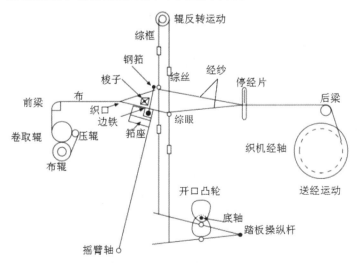

图 31.1　织机横截面简图

当任何经纱发生断裂时,停经片使织机停止。从这里开始,纱线先穿过综眼,再穿过钢筘的筘齿。按织物纹理升降综框的操作称为开口。在筘前,两片经纱与筘形成一个三角形的梭口。梭口形成后,携带纬纱的梭子横穿织物。这称为引纬。一根纬线被称为一纬。

2. 织造机理

织造机织物,织机需完成三个主要运动,即开口、引纬和打纬。除此之外,在织造中还有送经和卷取两个次要运动。在普通织机中,所有运动都由称为曲轴的主轴完成。曲轴旋转,一周内以不同的时间间隔停顿,完成织机的各运动。

3. 凸轮开口

开口机构由两个开口凸轮组成。平纹织造中,开口凸轮安装在底轴上,只需使用两个综框。但如需设计两个以上的综框,则需使用安装在底轴上的挺杆轴。开口凸轮安装在推杆轴上。

4. 多臂开口

多臂机构最多可以控制 30 个综框。有两种类型的多臂机构,消极式多臂和积极式多臂。在消极式多臂开口中,多臂提升综框,弹簧复位。在积极式多臂开口中,多臂控制综框升高和降低,去掉了弹簧。消极式或积极式多臂机又分为单动式提升和复动式提升。复动式多臂机的一个循环占两纬,因此它的大部分运动发生在织机提升半速时,可以更高速度运行。所有现代消极式多臂机均为复动式升降多臂机。虽然这种多臂机已大多被积极式多臂机所取代。

5. 提花开口

约瑟夫发明的提花机,是纺织行业中各种织机和针织机的原型。大且复杂的设计需使用提花开口织造。在这种开口中,经纱由综线单独控制,无综框。综丝和经纱数量一样多,这样可以编织出无限的图案。

6. 引纬

在梭织中,纬纱由梭子引入,梭子在织机上连续来回移动。织机两侧的打纬杆击打梭子,并使其飞过织机的开口来激活梭子。纬纱凸轮安装在织机的底轴上,并相互成 180 度。当打纬杆凸轮旋转时,它通过锥体移动打纬轴。这使得打纬轴通过吊带拉动打纬杆,这样打纬杆会加速并撞击梭子。开发了不同的打纬机理以适应织机的类型。

7. 送经运动

当纱线织造时,送经机构从织机经轴上退绕经纱,并控制送经量以维持最佳张力。如果经纱的张力没有控制在预设等级,经纱断头率会增加,这将影响织物的尺寸和物理性能。

8. 卷取运动

打纬后纲筘离开,织布由卷取运动带离织造区。为控制纬纱密度,卷取辊以一定速率带离布料,将其卷绕在织机的布辊上。根据所织织物的类型,卷取辊上覆盖有穿孔钢片或硬橡胶。系列齿轮机构控制卷取辊卷取纬纱的密度。目前,现代织机制造商使用电子式卷取,由伺服电机更好、更准确地控制纬纱间距。

Lesson 32　Modern weaving machines

For centuries woven fabrics used to be made in shuttle looms. The advent of new technologies led to the development of shuttleless looms which fulfilled the ever-increasing demand for better quality and higher productivity. In conventional shuttle looms, a shuttle weighing about 400g is used for inserting the weft yarn which weighs only less than 1/1,000th of the shuttle weight. The mechanism used to insert the weft in this system limits the speed of the loom to 200rpm–250rpm. As a result, methods of weft insertions without a shuttle have been developed. They are air jet, water jet, projectile and rapier.

In practice these looms are named after their weft insertion system. To distinguish from shuttle looms, these machines are called shuttleless looms or shuttleless weaving machines, which are considered to be the second generation of weaving machines. The first projectile weaving machine was introduced to the market in 1952. Production of rapier and air jet weaving machines started in 1972 and 1975 respectively.

1.Air jet weaving

Air jet weaving is a method of weaving in which a predetermined length of weft yarn is inserted into the warp shed by means of compressed air. The most popular configuration of air jet weaving is the multiple nozzle system and profiled reed. The method of weft insertion is shown in Fig. 32.1.

Yarn is drawn from the yarn package and stored in the accumulator. Due to high yarn velocity during insertion, it is difficult to unwind yarn intermittently from yarn package. Therefore, a yarn accumulator with feed systems is used between the tandem nozzle and yarn package. Yarn is released from the clamp of the accumulator as soon as the tandem and main nozzles are turned on. Upon release of the weft yarn from the clamp, it is fed into the warp shed via tandem and main nozzles. The combination of these two nozzles provides the initial acceleration for weft yarn to traverse across the warp shed. Subsequently, the sub or relay nozzles are activated to maintain the velocity of the leading end.

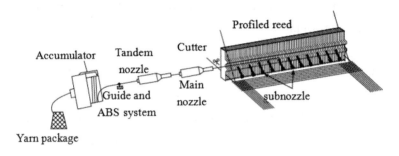

Figure 32.1   Air jet filling insertion with profiled reed

2.Water jet weaving

In water jet weaving machines the weft yarn is inserted by highly pressurized water. These looms are similar in many ways to air jet looms but they differ in construction, operating conditions and performance. Since water is used for weft insertion, warp and weft yarns must be water insensitive. The machine parts that get wet must be resistant to corrosion. The speed of weaving in this loom is high but the types of cloth that could be woven are somewhat limited due to the fact that yarn used for weaving should be wettable. These machines are commonly used for weaving synthetic filament yarns like polyester, nylon, etc.

3.Rapier weaving

In rapier weaving, a rigid or flexible rapier is used to insert the weft yarn across the warp sheet. The rapier head picks up the weft yarn and carries it through the shed. The rigid rapier is a metal bar generally with a circular cross-section. The flexible rapier has a tape-like structure that can be wound on a drum. Rapier weaving machines can be classified according to the method of weft insertion of single or double rapier system.

4.Projectile weaving

In a projectile weaving machine a projectile firmly holds the yarn by means of a gripper and traverses across the warp sheet. This system of filling insertion enables all yarns, whether fine or coarse, to be used as weft yarns. A variety of yarns made from cotton, wool, multifilaments, jute, and linen can be used as weft filling. This capability enables a variety of fabrics to be produced on these looms. The other important benefit in using a projectile weaving machine is that more than one width of fabric can be woven at a time.

5.Multiphase weaving

Rapid developments have been taking place in recent years in the application of newer technologies to weaving. Speeds of weaving machines along with its more sophisticated patterning capabilities play a major role in the present developments. Thus shuttle looms gave way to shuttleless weaving machines. Both shuttle looms and shuttleless looms are single phase weaving machines in which the shedding, weft insertion and beating-up operations form the sequence. The necessity of waiting to insert one pick after another limits the speed of single phase machines. In these machines, the weft insertion rate has reached a stagnation point of around 2,000m/min. Besides, the strain on the mechanisms employed and the stress on the yarns used for weaving have almost reached their optimum physical limits. All these factors led to the development of the multiphase weaving machine. In a multiphase weaving machine a number of weft yarns can be inserted simultaneously, unlike a single phase weaving machine where only one weft yarn at a time is placed in the fabric.

生词与词组

1.water jet 喷水式

2.projectile [prə'dʒektl]*n.* 片梭式

3.rapier ['reɪpiə(r)]*n.* 剑杆；剑杆式

4.predetermined length 预定长度

5.compressed air 压缩空气

6.configuration [kən,fɪgjə'reɪʃ(ə)n]*n.* 结构；布局；配置

7.nozzle ['nɒzl]*n.* 喷嘴

8.profiled reed 异形筘

9.yarn package 筒纱

10.tandem nozzle 串联喷嘴

11.main nozzle 主喷嘴

12.initial acceleration 初始加速度

13.traverse across 穿过

14.warp shed 经纱片

15.the sub or relay nozzle 辅助或中继喷嘴

16.pressurized ['preʃəraɪzd]*adj.* 加压的；受压的

17.resistant to corrosion 耐腐蚀

18.rigid or flexible rapicr 刚性或柔性剑杆

19.rapier head 剑杆头

20.tape-like 带状

21.drum [drʌm]*n.* 滚筒

22.gripper ['grɪpə]*n.* 夹子；夹持器

23.sophisticated patterning capability 复杂的花型

24.single phase machine 单相织机

25.stagnation [stægˈneɪʃ(ə)n]*n.* 停滞

26.optimum physical limit 最佳物理极限

译文

第 32 课　现代织机

梭织机织造梭织织物已有几个世纪了。为满足了高质量、高产的需求，新的无梭织机得以发展。传统的有梭织机，使用重约 400 克的梭引入纬纱，而纬纱的重量仅为梭子重量的 1/1000。该插入纬纱系统将织机的速度限制在 200–250 转 / 分。因此，无梭子的引纬方法发展开来。它们分别是喷气式、喷水式、片梭式和剑杆式。

事实上，这些织机均以其引纬系统而命名。为区分有梭织机，这些机器被称为无梭织机，其被称为第二代织机。1952 年第一台片梭织机投放市场。剑杆织机和喷气织机也分别于 1972 年和 1975 年开始生产。

1. 喷气织造

喷气织造是由压缩空气携带预定长度的纬纱引入经纱梭口的织造方法。多喷嘴系统和异形筘是喷气织造最佳的配置。喷气引纬方法如图 32.1 所示。

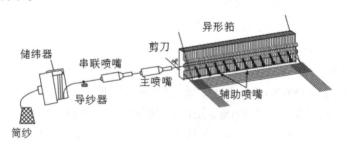

图 32.1　带异形筘的喷气织机引纬

纱线由筒纱中被抽出，并储存于储纱器。由于引纱速度高，从纱卷装中难以间歇退绕纱线。因此，在串联喷嘴和筒纱之间加装了带有喂入系统的储纱器。当主喷嘴和串联喷嘴运行时，纱线从蓄能器的夹具中释放出来。夹具释放纬纱，经串联喷嘴和主喷嘴进入经纱梭口。两喷嘴为纬纱穿过经纱梭口提供了初始加速度。随后,辅助喷嘴或中继喷嘴被激活以保持头端速度。

2. 喷水织造

在喷水织机中,纬纱由高压水引入。诸多方面,这类织机与喷气织机相似,但它们的结构、操作条件和性能均不同。因用水引纬,经纱和纬纱需对水不敏感。沾水机器部件需耐腐蚀。该织机的织造速度很高,但由于用于织造的纱线应该是可润湿的,因此可以织出的布的类型有所限制。该机器通常用于织造涤纶、尼龙等合成长丝纱线。

3. 剑杆织造

在剑杆织机中,刚性或柔性剑杆携带纬纱引入经纱片。剑杆头抓取纬纱并携带其穿过梭口。刚性剑杆一般为圆形截面的金属棒。柔性剑杆呈带状结构,其可缠绕在滚筒上。剑杆织机可按单剑杆系统或双剑杆系统的引纬方式分类。

4. 片梭织造

在片梭织机中,纱线由片梭的夹纱器握持,并由片梭携带横穿经纱片。该引纬系统适用于所有纱线,无论是细纱还是粗纱均可作为该系统的纬纱。棉、毛、丝、黄麻和亚麻制成的各种纱线均可作为该系统的纬纱。这使该织机可生产各种织物。片梭织机的另一个重要的好处是可织出多种宽度的织物。

5. 多向织造

近年来,在新技术应用方面织造技术得以快速发展。高速、复杂花型的织机是当前研发的主要方向。因此,有梭织机已被无梭织机超越。有梭织机和无梭织机都是单相织造机械,其开口、引纬和打纬过程依次进行。一纬一纬依次引入,限制了单相织机的速度。在这些织机中,引纬速度达到了 2000 米 / 分的停滞点。此外,织造机构的应变和纱线的应力已达到了最佳物理极限。所有这些因素促成了多相织机的发展。多相织机可同时插入多根纬纱,这与一次引入一根纬纱的单相织机有所不同。

Lesson 33 The development of knitting technology

Hand knitting is the foundation stone of today's mechanical and electronic stitch formation. In 1589, Lee invented the mechanical stitch formation technique on the stocking frame. His frame was able to knit 16 stitches at the same time. This technique is still used for today's modern machines. In 1758, J. Strutt invented the double knit technique, which consists of vertically arranged needles between the horizontal needles. In 1798, Decroix developed the circular knitting technique. In 1847, Townsend invented the latch needle, making stitch formation easier and increasing production speed. In 1878, Griswold invented the circular knitting machine, which can produce rib and plain fabric in any desired distribution by vertical cylinder and horizontal dial needles. In 1910, the interlock fabric was developed by the firm Scott and then in 1918, the first double cylinder, small circular knitting machine with double hook needle was developed by the firm Wildt. In the 1920s, mechanical needle selection devices such as punched tapes and pattern wheels began to be widely used. In the 1963, the first electronic needle selection with film-tape was demonstrated by the firm Morat at ITMA.

In the 1990s, the four needle bed technique with the flat knitting technology by Shima-Seiki started to be widely used, making it possible to produce a garment from a flat knitting machine without any sewing operation. The Mayer & Cie firm demonstrated that it is possible to make an intarsia technique on a circular knitting machine which has three to six times greater production capacity than a conventional flat knitting machine. With warp knitting technology, apart from conventional needle selection, it is now possible to individually select needles by the Piezo Jacquard system.

Stitch: stitch is the smallest unit in knitted fabric. A knitted fabric surface is formed by repeating it, side to side and one on top of the other (see Fig. 33.1). It consists of loop head, loop leg and loop feet.

Plain stitch: this is the technical face side of stitch where loop legs are above the neighbour stitch and loop head is below the neighbour stitch.

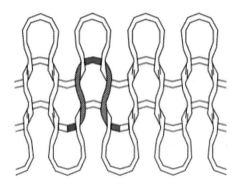

Figure 33.1   Knitted fabric surface

Classification of knitting technology

Knitting technology is classified into two main groups according to yarn presentation and yarn processing, i.e., weft knitting technology and warp knitting technology. In weft knitting technology, one yarn end is horizontally fed into all needles in the needle bed. Needles mostly are moved successively (individual needle motion as in circular knitting or V bed flat knitting) or simultaneously (collective needle motion as in straight bar knitting and loop wheel knitting). Yarn ends are mostly fed from a bobbin. Yarn ends in the fabric can easily be unravelled in a horizontal direction from the end knitted last. Products obtained from weft knitting technology are widely used in the apparel industry for pull-overs, t-shirts, sweatshirts, etc. Materials used are mostly natural fibres or blends with synthetic fibres. Weft knitting machines have relatively low investment cost, small floor space requirements, low stock holding requirements and quicker pattern change capabilities than warp knitting machines.

In warp knitting technology, one yarn end is longitudinally fed into one needle in the needle bed. Needles are moved simultaneously and collectively. Yarn ends are mostly fed from a warp beam. A yarn end in a fabric can hardly be unravelled in the vertical direction. Products obtained from warp knitting technology are widely used for household and technical textiles such as geotextiles, laces, curtains, automotive, swimwear, towelling, nets, sportswear, bed linen, etc. Materials used are mostly synthetic fibres in filament form. Higher machine speeds (up to 3,500rpm), finer gauges (up to 40 needles per inch), wider machines (up to 260 inches) and a multiaxial structure for technical application are available with warp knitting technology, compared to weft knitting technology. Generally, warp knit fabrics are less elastic than most weft knitted fabrics. They have a certain amount of elasticity in the width and a tendency to increase in a lengthwise direction after repeated wearing and washing.

Knitting technology developed from a very primitive method to high-tech methods, such as computer controlled production and design. During the last two decades, electronic control of knitting machines developed rapidly. Different fibre types such as cotton, wool, polyester, acrylic, viscon, etc., can be used on knitting machines. However, cotton and cotton blend yarn are widely used on circular knitting machines for the apparel industry producing t-shirts, sweatshirts and underwear, due to their comfort and healthy properties. Although different man-made fibres have been developed, none of them could still attain the properties of cotton. It seems therefore that cotton and cotton blend yarn will be preferred for use in sportswear and underwear, for a long time.

生词与词组

1.hand knitting 手工编织
2.stitch formation 针织成形

3.mechanical stitch formation technique 机械式针织成形技术

4.stocking frame 袜架

5.horizontal needle 平针

6.circular knitting technique 圆形针织技术

7.rib and plain fabric 罗纹和平纹织物

8.vertical cylinder 垂直圆筒

9.horizontal dial needle 水平针盘针

10.interlock fabric 锁链织物

11.double cylinder 双圆筒

12.double hook needle 双钩针

13.mechanical needle selection device 机械选针装置

14.punched tape and pattern wheel 打孔带和花轮

15.four needle bed technique 四针床技术

16.flat knitting technology 横机技术

17.intarsia technique 嵌花技术

18.weft knitting 纬编

19.warp knitting 经编

20.circular knitting 圆织

21.V bed flat knitting V 形床横机

22.straight bar knitting 直条编织

23.loop wheel knitting 环轮编织

24.bobbin [ˈbɒbɪn]*n*. 线轴

25.pullover [ˈpʊləʊvə(r)]*n*. 套头衫

26.stock holding requirement 库存要求

27.pattern change capability 花样更换能力

28.longitudinally [ˌlɒndʒɪˈtjuːdɪnəli]*adv*. 纵向地,长度上地

29.geotextile [dʒiːəʊˈtekstaɪl]*n*. 土工布

30.toweling [ˈtaʊəlɪŋ]*n*. 毛巾

31.bed linen 床上用品

32.finer gauge 针距更细

33.lengthwise [ˈleŋθwaɪz]*adj*. 纵向的

译文

第33课　针织技术发展

手工编织是当今机械和电子针编成形技术的基石。1589年,李(Lee)在长袜框架上编织,发明了的机械针织成形技术。他的框架可同时编织16针。该技术仍然为现代机器所用。1758年,斯克鲁特(Strutt)发明了双面针织技术,它由水平针和垂直针组成。1798年,德克鲁瓦(Decroix)开发了圆形针织技术。1847年,汤森德(Townsend)发明了舌针,其更易针织线迹成形,并提高了生产速度。1878年,格里斯沃尔德(Griswold)发明了圆形针织机,它由垂直圆筒和水平针盘针组成,可生产任意的罗纹和平纹织物。1910年,斯科特(Scott)公司开发了锁链织物。1918年,维尔特(Wildt)公司开发了第一台双圆筒、双钩针小圆机。1920年,打孔带、花轮等机械选针装置开始广泛使用。1963年,莫拉特(Morat)公司在ITMA(国际纺织机械展览)展示了第一台带有薄膜胶带的电子选针器。

20世纪90年代,来自岛津的四针床横机技术推广应用,使得无缝纫操作编织服装由横机可实现。迈耶·西(Mayer & Cie)公司发明了可进行起花操作的针织圆机,且该机的生产力为传统横机的3-6倍。除了传统的选针外,经编技术现已可用压电提花系统单独选针。

线圈:线圈是针织物的最小单位。针织物是由线圈从一侧到另一侧,从一个向上至另一个,重复而形成的(见图33.1)。它由圈弧、圈柱和沉降弧组成。

平针:这是线圈的工艺面,其圈柱在相邻线圈上方,圈弧在相邻线圈下方。

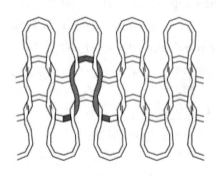

图33.1　针织物表面

针织技术分类

针织技术根据纱线穿插和加工形式分为纬编和经编两大类。纬编中,一根纱水平喂入针床中的所有织针。织针多为连续运动(单针运动,如圆机或 V 形床横机)或同步运动(集体针运动,如直条编织和环轮编织)。纱由线轴喂入。纱头可从编织物的末端沿水平方向脱散。服装也广泛应用针织技术的织物,如套头衫、T 恤、运动衫等。针织物的原料主要为天然纤维或天然纤维与合成纤维的混纺织物。与经编机相比,纬编机具有较低的成本、较小的占地面积、较低的库存要求和更快的花样更换能力。

经编技术是纱线沿纵向喂入针床的针,各针同步移动,纱线沿经轴喂入。织物的纱线不易沿经向脱散。经编产品广泛用于土工布、花边、窗帘、汽车内饰、泳装、毛巾、网、运动服、床上用品等家用和产业用纺织品。经编原料多为合成纤维长丝。与纬编技术相比,经编技术的机器速度更高(可达 3500 转 / 分),针距更细(每英寸可达 40 针),机器更宽(可达 26 英寸)以及可用于多轴编织。通常,经编针织物的弹性比纬编针织物的弹性小。经编织物的横向具有一定弹性,其经反复穿着和洗涤后长度有所增加。

针织技术正从非常原始方式向高技术转变,如计算机控制的生产和设计。以往二十年,针织机的电控系统发展迅速。各种纤维均可应用于针织机,如棉、羊毛、涤纶、腈纶、黏胶纤维等。基于舒适和健康考虑,棉和棉混纺纱被圆机针织机织造为 T 恤、运动衫和内衣等服装。尽管研发了各种人造纤维,但人造纤维的性能仍差于棉。因此,在相当长一段时间内,棉和棉混纺纱将成为运动服和内衣的首选。

Lesson 34   Weft knitting technology

There are several different types of weft knitting machines, such as circular knitting machines (large diameter), hosiery machines, V bed flat knitting machines, flat bed purl machines, etc. However, especially for outerwear and underwear, there are two main widely used weft knitting machines, i.e., circular knitting machines (large diameter) and V bed flat knitting machines. The main differences between the circular knitting machines (large diameter) and V bed flat knitting machines are described below:

The machine frame of the circular knitting machine is of a cylindrical shape, while the machine frame of the V bed flat knitting machine is flat with the needle beds located approximately 90 degrees to each other.

Products of the circular knitting machines have a tubular form, while flat form fabrics are produced by the V bed flat knitting machines.

On V bed flat knitting machines, the needle bed is stationary and the carriage, which contains a cam system, traverses the machine width and activates the needles. In circular knitting, the needle bed generally rotates and stationary cam blocks around the needle bed activate the needles. Production rates in circular knitting are generally higher than flat knitting, due to higher velocity and higher number of knitting (cam) systems.

Flat knitting machines have more versatility than circular knitting machines. More complex fabric designs can be produced by flat knitting machines.

Flat knitting machines have generally a coarser machine gauge than circular knitting machines, thus coarser fabric structures used mainly for outerwear in cold weather conditions such as pullovers, sweaters, etc., are produced.

1.Circular knitting machines

In circular knitting (Fig. 34.1), the machine gauges and the machine diameters range from 4 to 32 needles per inch and 8 to 48 inches in diameter, respectively. They can run at the speed of 76rpm and may have up to 132 yarn feeders.

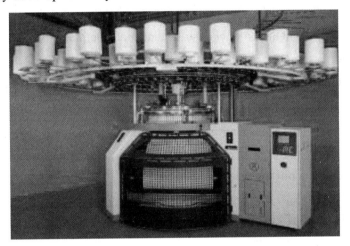

Figure 34.1   Circular knitting machine

The machine frame of a circular knitting machine is of cylindrical form and fabric is produced in tubular form. Figure 34.2 show the arrangement of main machine parts of a plain circular knitting machine used for the production of single jersey fabrics. In most circular knitting machines the needle beds rotate while the cam blocks are stationary. During rotation, the needle butt contacts with the cam and produces a knit or a tuck stitch. At every knitting system (cam together with yarn feeder) on a circular machine frame, the needle can produce a knit, tuck or float stitch.

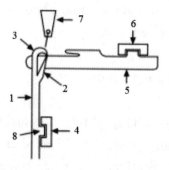

1.Needle; 2.Needle latch; 3.Needle head; 4.Needle cam block; 5.Holding down knocking over sinker; 6.Sinker cam block; 7.Feeder; 8.Needle butt

Figure 34.2　Illustration of circular knitting machine(plain circular knitting machine)

Thus, for example, if a plain knitting machine has a 90 knitting system, at the end of one revolution, one needle can produce 90 stitches (90 courses). If the knitting machine is interlock type (1 × 1 interlock), two knitting systems produce one course, thus, at the end of one revolution, 45 courses are produced.

2.V-bed flat knitting

Machine gauge in the flat knitting (see Fig. 34.3) is in the range of 3npi and 18npi (needles per inch). Needle bed widths are in the range of 30cm and 244cm or more. Number of cam systems (cam together with feeder) at one side of one cam carrier is not more than four. Average knitting speed is around 1.2 metres per second. In contrast to the circular knitting machines, the machine frame of flat knitting machines is in flat form, and knitted fabrics are also in flat form. Nowadays, most flat knitting machines are electronically controlled. The needle bed is in V form and the angle between rear and front needle bed is about 90 and 104 degrees to each other. Needles at the front and rear needle beds are arranged in a rib setting. The needle bed is stationary while the carriage that contains the cam systems (cam and feeder) travels along the needle bed. The yarn feeder is set on the rails that extend across the width of the machine.

During traversing, the carriage takes a yarn feeder on the rails and moves along the needle bed with it. While the carriage moves along the needle bed, the needle butt contacts with the cam on the carriage and produces stitch, tuck stitch or float. The yarn that is taken over the cone continues through the knot catcher assembly and spring loaded brake disc. It then passes through the tension arm. A brake disc is used to tension the yarn between cones and yarn feeder. The tension arm is used to pull out excess yarn when the carriage traverses. Flat fabric that is not produced in tubular form, as in a circular knitting machine, is pulled down by take-down rollers.

Figure 34.3   V bed flat knitting machine

Modern electronic V bed flat knitting machines possess a holding down sinker placed at the front edge of the needle bed between needles. It is used to hold the fabric when the needle is raised into the clearing position. The selection of the needle for knit, tuck or float stitch or transferring/receiving position is carried out by an electronic system.

生词与词组

1.hosiery machine 袜机

2.flat bed purl machine 平床绗缝机

3.carriage ['kærɪdʒ] *n.* 机头

4.single jersey fabric 单面针织织物；汗布

5.cam [kæm] *n.* 三角

6.tuck stitch 集圈组织

7.needle latch 针舌

8.cam block 三角片

9.gauge [geɪdʒ] *n.* 轨距；隔距

10.needle butt 针踵

11.knot catcher 纱结器；除结器

12.brake disc 盘式制动器

13.tension arm 张力杆

14.take-down roller 牵拉辊

15.holding down sinker 握持沉降片

译文

第34课 纬编技术

纬编机包括圆机（大直径）、袜机、V床横机、平床绗缝机等。然而，圆机（大直径）和V床横机是使用较多的两类纬编机，特别适用于外套和内衣。圆机（大直径）与V形床横机的主要区别如下：

（1）圆机的机架是圆柱形的，V形床横机的机架是直的，前后针床相互成90度。

（2）圆机的织物呈管状，V形床横机的织物呈片状。

（3）V形床横机的针床是固定的，机头（含三角系统）沿机床移动并推动织针。圆机的针床旋转，针床固定的三角带动织针。圆机具有更高的速度和更多的三角系统，其生产率高于横机。

（4）横机比圆机更通用。复杂的织物可由横机设计生产。

（5）横机的轨距比圆机大，其生产的织物结构更粗糙，多用于织造御寒外套，如套头衫、毛衣等。

1. 圆机

圆机（见图 34.1）的机器轨距为每英寸 4 至 32 针，机器直径为 8 至 48 英寸。圆机的运行速度可达 76 转／分，最多可携带 132 个纱嘴。

图 34.1　圆机

圆机的机架为圆形，织制的织物呈管状。图 34.2 为织制单面针织物的单面圆机的主要部件。多数圆机为针床旋转、三角静止。在旋转过程中，针踵与三角接触并产生编织或集圈的针迹。圆机的编织系统（三角与纱嘴）引导织针进行编织、集圈或浮针。

因此，如平针织机有 90 针编织系统，则其运转转一圈，每一枚针可编织 90 针（即 90 个横列线圈）。如果针织机是罗纹型（1×1 互锁罗纹），则两个针织系统编织一个横列，运转转一圈可生产 45 个横列。

2. V 形床横机

横机（如图 34.3）中的机针轨距为 3-18 针（每英寸针数）。针床宽度为 30-244 厘米或以上。三角系统（三角和喂纱器）的数量一般不超过 4 个。平均编织速度约为 1.2 米／秒。与圆机不同，横机机架是直的，编织织物为片状。目前，多数横机均为电子控制。针床呈 V 字形，前后针床夹角约为90度和104度。前后针床的机针呈交叉排列。针床是固定的,机头携带三角系统(三角和喂纱器)沿针床移动。喂纱器安装在与机器等宽的滑轨上。

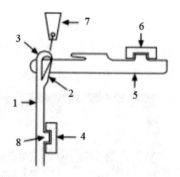

1.针;2.针舌;3.针头;4.针三角;5.握持脱圈沉降片(生克);6.生克三角片;7.纱嘴;8.针踵

图 34.2　圆机图例(单面圆机)

图 34.3　V形床横机

机头滑动期间,导轨上的喂纱器也随之沿针床移动。机头沿针床移动,针踵与滑轨上的三角接触并发生编织、集圈或浮线线迹。引入纱线穿过纱结器组件和弹簧制动盘,再穿过张力杆,张力盘用于调节纱结器组件和喂纱嘴之间的纱线,张力杆用于调节机头横动带出的多余纱线。在圆机中,非管状的扁平织物由牵拉辊牵拉。

　　现代电子 V 形床横机设有一握持沉降片位,装在针床前缘与针之间,其用于织针升起脱圈时下压织物。电子系统用于针织、集圈或浮针或传送 / 接收位置的针的选择。

Lesson 35 Warp knitting technology

There are two main warp knitting machine (see Fig. 35.1) classifications, Tricot and Raschel. The main differences between them are described below.

Figure 35.1 Warp knitting machine

Raschel machines are generally coarser and slower than Tricot machines due to a higher number of guide bars (up to 80 guide bars for Raschel compared with up to 4 guide bars for Tricot) and have a longer and slower needle movement.

Raschel machines are much more versatile, i.e., most types of yarns and slit films can be used. Complex fabric designs can be produced on Raschel machines, while Tricot machines are limited to only basic fabric designs.

The sinker used on a Tricot machine controls the fabric and holds the fabric while the needles rise to the clear position. However, in Raschel knitting, the fabric is controlled by a high take-down tension and the sinker is not so important for fabric control. The fabric produced on a Raschel machine is pulled tightly downwards (about 160 degrees), while fabric produced on a Tricot machine is pulled gently from the knitting zone (about 90 degrees).

The fabric take-up mechanism in Tricot machines is positioned further away from the knitting zone, however, in Raschel machines, it is positioned close to the knitting zone.

In the past, bearded needles and latch needles were used for Tricot and Raschel machines, respectively. However, nowadays, compound needles have replaced the bearded and latch needle in warp knitting technology.

While the machine gauge in Tricot is described as the number of needles per inch, it is generally described in Raschel machine as the number of needles per two inches.

The chain links in Tricot machines are numbered as 0, 1, 2, 3, 4, etc., while the chain links in Raschel machines are numbered in even numbers, such as 0, 2, 4, 6, etc. Tricot machine gauge is in the range of 6–44 and knitting width can reach up to 260 inches.

生词与词组

1.guide bar 梳栉；导纱梳栉；导杆
2.take-up mechanism 卷取机构；卷取装置
3.Tricot machine 特里科机
4.Raschel machine 拉舍尔机
5.bearded needle 钩针
6.latch needle 舌针

译文

第 35 课　经编技术

经编机主要分为特里科和拉舍尔两类(见图 35.1)。它们之间的主要区别如下。

图 35.1　经编机

因梳栉数量较多(拉舍尔机最多 80 把梳栉,而特里科机最多 4 把梳栉),通常拉舍尔经编机的规格比特里科经编机更大,针运行速度也更慢。

拉舍尔经编机的用途更广,大多数的纱线和切膜均适用。复杂织物可由拉舍尔经编机设计生产,而特里科机仅限于基本组织织物设计。

当织针上升至清纱位置时,特里科机的沉降片用于控制、握持织物。然而,拉舍尔机由高拉力的张紧器对织物进行控制,沉降片对织物控制较弱。拉舍尔机编织的织物需从编织区向下拉紧(约 160 度),而特里科机编织的织物需从编织区轻轻拉出(约 90 度)。

特里科机的织物卷取机构离编织口较远,而拉舍尔机的织物卷取机构靠近编织口。

以往，特里科机和拉舍尔机使用钩针和舌针。在当前经编技术中，复合针已经取代了钩针和舌针。

特里科机的机器轨距以每英寸的针数计量，而拉舍尔机通常以每两英寸的针数计量。

特里科机的链节编号为 0、1、2、3、4 等，而拉舍尔机的链节编号为偶数，如 0、2、4、6 等。特里科机的轨距为 6-44，编织宽度可达 260 英寸。

Lesson 36 Design in textiles

Every textile product is designed: that is, it is made specifically to some kind of plan. Design decisions are made at every stage in the manufacturing process—what fibres should be used in a yarn, what yarns in a fabric, what weight of fabric should be produced, what colours should the yarn or fabric be produced in, what fabric structures should be used and what finishes applied. These decisions may be made by engineers and technologists in the case of industrial or medical textiles where performance requirements are paramount, or, more often in the case of apparel, furnishings and household textiles, by designers trained in aesthetics, technology and marketing. The designers found in the textile and clothing industries are frequently involved throughout the design process, from initial identification of a need/requirement, through research, generation of initial design ideas, design development and testing to ultimate product specification.

1.The principles and elements of textile design

Textiles are frequently made to be decorative and are used to embellish and decorate both people and objects. The designers responsible for such textiles have to balance many factors when answering a design brief. What is the fabric for? How must it perform? Who is the customer? What are the economic restraints? How is the fabric to be produced? etc. Research into how people select products shows that colour and appearance are two of the most significant factors, with handle, performance and price coming lower down in terms of importance. Textile designers therefore need to have a good understanding and sensitivity to colour and aesthetics.

2.The diversity of textile design

The diversity of the textile and clothing industries is reflected in the many different types of designers needed. At every stage of manufacture of textiles there are colourists determining the fashion colours in which the fibres will be produced, yarn designers developing yarns to meet certain requirements, knitted-fabric designers, woven-fabric designers, carpet designers, print designers, embroidery designers, knitwear designers, designers of women's wear, men's wear and children's wear, accessory designers, designers of casualwear, sportswear, eveningwear, swimwear and designers for the mass market, haute couture and designer labels, etc.

3.Printed

Textile designers may be categorised by the types of product or fabrics for which they design. Printed textiles are often considered to include fabrics patterned by dyeing techniques as well as those where the design is applied to the fabric by a printing process.

The overall purpose of a textile designer is to design and produce to an agreed timetable, an agreed number of commercially viable textile designs.

生词与词组

1.furnishings [ˈfɜːnɪʃɪŋz]n. 家具；装饰品

2.initial identification 初步鉴定

3.initial design idea 最初的设计理念

4.colourist [ˈkʌlərɪst]n. 调色师；配色技师

5.embroidery [ɪmˈbrɔɪdəri]n. 刺绣；绣花

6.casualwear [ˈkæʒuəlˌweə]n. 休闲装；便装

7.eveningwear [ˈiːvnɪŋˌweə]n. 晚礼服；晚装
8.haute couture 高级时装
9.agreed timetable 既定时间表

译文

第 36 课　纺织品设计

任何纺织产品都需要设计：也就是说，它是依据特定计划而制造的。每一阶段的加工均需设计决策：纱线使用何种纤维，织物使用何种纱线，生产多重的织物，纱线或织物染何种颜色，采用何种结构的织物以及何种表面整理。对于性能要求严格的工业或医用纺织品，这些决定由工程师和技术人员做出，或者，在服装、家具和家用纺织品的情况下，其通常由受过美学、技术和营销培训的设计师做出。从需求／要求的拟定，到研究、生成初步的设计理念、设计试制以及测试最终的产品，纺织和服装行业的设计师经常参与整个设计过程。

1. 纺织品设计的原则和要素

纺织品经常被制作成装饰品，用于点缀或装饰人和物体。负责纺织品的设计师在填写设计计划表时需平衡各种因素。面料如何用？它如何展现？谁是客户？经济上的限制有什么？面料如何生产？等等。对人们如何选择产品的研究表明颜色和外观是两个最重要的因素，而手感、性能和价格的重要性则较低。因此，纺织品设计师需要对色彩和美学有深入的理解和敏感性。

2. 纺织品设计的多样性

纺织和服装行业的多样性体现在对各类设计师的需求上。每阶段纺织品生产都需调色师确定纤维的流行颜色，纱线设计师按照特定需求设计纱线，针织面料设计师，机织面料设计师，地毯设计师，印花设计师，刺绣设计师，针织品设计师，女装、男装和童装设计师，配饰设计师，休闲装、运动装、晚装、泳装和大众市场设计师、高级时装和品牌设计师等。

3. 印花

纺织品设计师可根据他们设计的产品或面料的类型对其进行分类。通常印花纺织品包括染色图案的织物以及印花工艺设计的织物。纺织品设计师的总体目标是按照既定时间表设计和生产既定数量的商业上可行的纺织品。

Lesson 37　Jacquard textile design

Jacquard textile weaving is an ancient craft with a century-old history. The design and production of jacquard textiles has always been regarded as tedious and time-consuming endeavours, in which considerable skill and experience are required to produce hand-drawn patterns to form figured woven fabric. Due to the intricacy and unique design of the woven colours and patterns, jacquard textiles and related jacquard products have extended their applications to a wide range of fashion materials, home furnishings and decorations. Theoretically, a basic difference exists in the effect of colour and pattern between jacquard woven fabrics and printed fabrics. For printed fabrics, the printed pattern is a result of superimposing several transparent colour inks.

Such superimposition of transparent inks enables a pattern to be reproduced on fabric with over a million shades. Hence, the pattern reproduced is very close to its original. For jacquard fabric as well as for woven fabric, however, pattern is reproduced through a kind of woven figuration, where the colour and pattern effect are dependent on the woven structure of interlacing warp ends and weft picks. Due to the different colour theories and restrictions of woven structures, the pattern of jacquard fabric ought to be designed with reference to the weaving and figuring technical conditions, such as fabric density, materials, and the manner of mounting of the jacquard machine. In addition, since the structure design of jacquard fabric is approached in a traditional single-plane design mode, and a one-to-one corresponding principle, i.e., designing weave one by one according to the effect of each colour drawn on a certain pattern, currently the colour expression of jacquard fabric is limited to not more than 100 colours in each pattern design.

Even by employing CAD systems today, the design principles and processes are still subjected to a plane design mode. Thus, the colour and pattern effects of jacquard fabric remain very much the same in terms of expression. It is still a major challenge for jacquard textile designers to be able to design a method that enables the creation of print-like patterns on figured woven fabrics that can be processed and produced conveniently on an industrial scale.

Capitalising on digitisation technology, digital image design features higher efficiency in design processing and greater compatibility in design applications. The effects produced by digital images can be more picturesque and imaginative than those expressed freehand. It was therefore only a matter of time before digital design technology brought innovation to the design and production of jacquard textiles. Over the past ten years, for example, research had been carried out to study computer-aided design via digitisation technology, with the purpose of enhancing the design efficiency of jacquard fabric.

However, because of the unresolved constraints of plane design mode, the design of jacquard textiles has remained unchanged and digital image design, via CAD, has been employed only to replace hand-drawn patterns; digital technology was not directly applied to the creation of jacquard textile designs. Since the structural design of the fabric plays the most important role in the creation of jacquard textiles, an attempt has now been made to bring innovation to the traditional principles and methods of structural design through the deployment of digitisation technology.

In addition to structure design, the colour theory of woven fabric is another important factor in the innovation of jacquard fabric. For jacquard fabric, as well as for woven fabrics constructed with opaque colour threads, the resultant colour effect exhibited on the face of the fabric is subject to optical colour mixing. By tradition, jacquard fabric design is a mechanical reproduction under a single-plane design mode that aims to imitate the colour and pattern effects of hand paintings. The potential aesthetic innovation of colour and figuration of the woven structure of the fabric have largely been overlooked and underexplored.

生词与词组

1.jacquard textile weaving 提花织物织造

2.ancient craft 古老技艺

3.time-consuming endeavour 费时费力

4.hand-drawn pattern 手绘图案

5.intricacy [ˈɪntrɪkəsi]*n.* 错综复杂

6.unique design 独特的设计

7.home furnishings 家居用品

8.superimposition of transparent ink 透明色颜料叠加 / 复配

9.interlacing warp end 交织的经纱末端

10.weaving and figuring technical condition 织造计算技术条件；织造设计工艺

11.single-plane 单面

12.one-to-one corresponding principle 一一对应的原则

13.colour expression 色彩表现；色彩表达

14.print-like pattern 印花图案

15.digitisation technology 数字化技术

16.picturesque and imaginative 多彩且富有想象力的

17.unresolved constraint 未解决的局限；未解析的约束

18.opaque colour thread 不透明色线

19.optical colour mixing 光色混合；混色

20.mechanical reproduction 机械复制；机械式复制

译文

第 37 课　提花织物设计

提花织物织造是一项有着百年历史的古老工艺。提花织物的设计和生产被认为是一项烦琐而耗时的工作,其需要相当多的技能和经验来制作手绘图案以形成花样机织物。因其复杂、独特的设计颜色和图案,提花织物及其相关产品已应用于时尚材料、家居装饰和装饰品。理论上,在颜色和图案的效果方面,提花织物和印花织物存在着根本的区别。对于印花织物,印花图案是几种透明色颜料叠加的效果。

这种叠加效果的透明油墨可在织物上再现一百万种色调。因此,再现图案非常接近其原始图案。对于提花织物和机织织物,图案是通过一种机织结构再现的,其中颜色和图案效果取决于交织的经纱和纬纱的编织结构。由于色彩理论的差异和机织结构的限制,提花织物的花型应按照织物密度、材料、提花的安装方式等织造工艺条件进行设计。另外,由于提花面料的结构设计采用一一对应的传统平面设计模式,即根据每种颜色在图案上绘制的效果逐一设计编织,目前,提花面料每个图案设计的色彩表现限制在 100 种颜色以内。

即使当今的 CAD 系统,其设计原则和过程仍受制于平面设计模式。因此,提花织物的颜色和图案所展现的效果相近。对于提花纺织品设计师来说,设计一种类似印花的图案,并可工业化生产的提花织物,仍是一大挑战。

利用数字化技术,数字图像设计效率更高,兼容性更好。数字图像的表达效果比徒手表达更多彩且更富有想象力。因此,数字设计技术带动提花纺织品设计和生产的变革,只是时间问题。例如,过去的十年,为了提高提花织物的设计效率,已展开计算机辅助数字化设计的研究。

然而,因平面设计模式的限制未解决,提花纺织品的设计未发生变化,仅通过 CAD 数字图像设计来替代手绘图案;数字技术并未直接应用于提花纺织品设计的创造。在提花纺织品的创造中,面料的结构设计至关重要,因此现在已尝试通过数字化技术来改变传统的结构设计原则和方法。

除了结构设计,机织面料的色彩理论是提花面料变革的另一个重要因素。提花织物及不透明色线编织的织物,织物表面所呈现颜色效果会受混色效果影响。传统提花面料设计是一种单面模式的机械复制,旨在模仿手绘的色彩和图案效果。织物组织结构和色彩的创新被忽视,仍未被充分探索。

Lesson 38　Nonwovens

1.Definition and classification

The term "nonwoven" arises from more than half a century ago when nonwovens were often regarded as low-price substitutes for traditional textiles and were generally made from drylaid carded webs using converted textile processing machinery. The yarn spinning stage is omitted in the nonwoven processing of staple fibres, while bonding (consolidation) of the web by various methods, chemical, mechanical or thermal, replaces the weaving (or knitting) of yarns in traditional textiles.

Therefore, the nonwoven industry as we know it today has grown from development in the textile, paper and polymer processing industries. The nonwoven industry is reluctant to be associated with the conventional textile industry and its commodity associations, nor would it want its products to be called "nonpapers" or "nonplastics".

The illusion created by this misnomer has been for some to think of nonwovens as some kind of bulk commodity, even cheap trade goods, when the opposite is often true. EDANA (The European Disposables and Nonwovens Association) defines a nonwoven as "a manufactured sheet, web or batt of directionally or randomly orientated fibres". To distinguish wetlaid nonwovens from wetlaid paper materials, the following differentiation is made, "more than 50% by mass of its fibrous content is made up of fibres with a length to diameter ratio greater than 300". INDA, North America's Association of the Nonwoven Fabrics Industry, describes nonwoven fabrics as sheet or web structures bonded together by entangling fibres or filaments, by various mechanical, thermal and/or chemical processes. Nonwovens are engineered fabrics that can form products that are disposable, for single or short-term use or durable, with a long life, depending on the application. Nonwoven technology also exists to approximate the appearance, texture and strength of conventional woven and textile fabrics and in addition to flat monolithic fabrics, multi-layer nonwoven composites, laminates and three-dimensional nonwoven fabrics are commercially produced. The most common products made with nonwovens listed by INDA include:

(1) disposable nappies;

(2) sanitary napkins and tampons;

(3) sterile wraps, caps, gowns, masks and curtains used in the medical field;

(4) household and personal wipes;

(5) apparel interlinings;

(6) carpeting and upholstery fabrics, padding and backing;

(7) agricultural coverings and seed strips;

(8) automotive headliners and upholstery;

(9) civil engineering fabrics/geotextiles.

2. Raw materials

Man-made fibres completely dominate nonwoven production, accounting for over 90% of total output. Man-made fibres fall into three classes, those made from natural polymers, those made from synthetic polymers and those made from inorganic materials. The world usage of fibres in nonwovens production is polypropylene 63%, polyester 23%, viscose rayon 8%, acrylic 2%, polyamide 1.5%, other speciality fibres 3%. Polypropylene fibres are predominant in the nonwoven industry. The share of viscose rayon is thought to have increased due to its increased importance in the spunlace wipe market. The solvent spun cellulosic fibre, Lyocell is becoming increasingly important in the nonwoven industry partly as a result of its absorbency and high wet strength.

Web formation

In all nonwoven web formation processes, fibres or filaments are either deposited onto a forming surface to form a web or are condensed into a web and fed to a conveyor surface. The conditions at this stage can be dry, wet, or molten-drylaid, wetlaid or polymer-laid (also referred to as spunlaid and spunmelt processes). Web formation involves converting staple fibres or filaments into a two-dimensional (web) or a three-dimensional web assembly (batt), which is the precursor for the final fabric. Their structure and composition strongly influences the dimensions, structure and properties of the final fabric. The fibre orientation in the web is controlled during the process using machinery adapted from the textile, paper or polymer extrusion industries. The arrangement of fibres in the web, specifically the fibre orientation, governs the isotropy of fabric properties and most nonwovens are anisotropic. Although it is possible to make direct measurements of the fibre orientation in a web, the normal approach is to measure the machine direction/cross direction (MD:CD) ratio of the web or more usually the fabric.

This ratio of fabric properties, usually tensile strength, measured in the machine direction (MD) and cross direction (CD) reflects the fibre orientation in the fabric. Commercially, obtaining a web or a fabric with a truly isotropic structure, that is, with an MD:CD=1, is rarely achieved and technically is frequently unnecessary. Other critical fabric parameters influenced at the web formation stage are the unfinished product weight and the manufactured width. Traditionally, each web-forming system was used for specific fibres or products, although it is increasingly common for similar commercial products to be made with different web formation systems. One example is in the manufacture of highloft nonwovens which can be produced with either a card and crosslapper or a roller-based airlaid system. In the hygiene industry, there is an increasing preference for the soft, staple fibre products produced by carding and hydroentanglement in favour of the alternative airlaid and thermal bonded products.

4.Web bonding

Nonwoven bonding processes can be mechanical, chemical (including bonding using solvents) or thermal. The degree of bonding is a primary factor in determining fabric mechanical properties (particularly strength), porosity, flexibility, softness, and density (loft, thickness). Bonding may be carried out as a separate and distinct operation, but is generally carried out in line with web formation. In some fabric constructions, more than one bonding process is used. Mechanical consolidation methods include needle punching, stitch bonding, and hydroentangling. In respect of needle punching, which is most commonly fed by a card and crosslapper, the world production is in excess of an estimated 1.1 million tonnes of needle felts of which over 72% used new fibres as opposed to recycled fibres. Chemical bonding methods involve applying adhesive binders to webs by spraying, printing, or foaming techniques. Solvent bonding involves softening or partially solvating fibre surfaces with an appropriate chemical to provide self- or autogeneously bonded fibres at the cross-over points. Thermal bonding involves the use of heat and often pressure to soften and then fuse or weld fibres together without inducing melting.

5.Drylaid nonwovens

The first drylaid systems owe much to the felting process known since medieval times. In the pressed felt industry, cards and web lappers were used to make a batt containing wool or a wool blend that is subsequently felted using moisture, agitation and heat. Some of the drylaid web forming technologies used in the nonwoven industry, specifically carding, originate from the textile industry and manipulate fibres in the dry state. In drylaid web formation, fibres are carded (including carding and cross-lapping) or aerodynamically formed (airlaid) and then bonded by mechanical, chemical or thermal methods. These methods are needlepunching, hydroentanglement, stitchbonding (mechanical), thermal bonding and chemical bonding.

6.Wetlaid nonwovens

Paper-like nonwoven fabrics are manufactured with machinery designed to manipulate short fibres suspended in liquid and are referred to as "wetlaid". The use of the wetlaid process is confined to a small number of companies, being extremely capital intensive and utilising substantial volumes of water. In addition to cellulose papers, technical papers composed of high performance fibres such as aramids, glass and ceramics are produced.

生词与词组

1.low-price substitute 低价替代品；低价代用品

2.drylaid carded web 干法梳理网

3.commodity association 商品协会

4.nonplastic [ˌnɒnˈplæstɪk]*adj.* 非塑料的

5.misnomer [ˌmɪsˈnəʊmə(r)]*n.* 误称；用词不当

6.bulk commodity 大宗商品

7.cheap trade good 廉价贸易商品

8.European Disposables and Nonwovens Association 欧洲一次性用品和非织造布协会

9.directionally or randomly orientated fibre 定向或随机取向的纤维

10.wetlaid nonwoven 湿法无纺布

11.entangling fibre or filament 缠结的纤维或长丝

12.flat monolithic fabric 单层布

13.laminate [ˈlæmɪnət]*n.* 层压材料

14.disposable nappy 一次性尿布

15.sanitary napkin 卫生巾

16.tampon [ˈtæmpɒn]*n*. 棉球；棉条

17.sterile wrap 无菌包布；无菌包扎

18.apparel interlining 服装衬布

19.automotive headliner 汽车顶篷

20.spunlace wipe 水刺擦拭布；水刺无纺布

21.molten-drylaid 熔融干法纺丝

22.spunmelt [ˈspʌnˌmelt]*n*. 纺熔法；纺粘与熔喷法

23.precursor [priˈkɜːsə(r)]*n*. 前道，先驱

24.anisotropic [ˌænaɪsəˈtrɒpɪk]*adj*. 各向异性的

25.machine direction 机器方向；纵向；纤维方向

26.cross direction 交叉方向；横向

27.highloft nonwoven 高蓬松无纺布

28.crosslapper [krɔːs ˈlæpə]*n*. 交叉铺网机

29.roller-based airlaid 辊式气流成网

30.hygiene [ˈhaɪdʒiːn]*n*. 卫生；卫生学

31.hydroentanglement [ˌhaɪdrəʊɪnˈtæŋglmənt]*n*. 水刺法

32.thermal bonded 热黏合的

33.needle punching 针刺法

34.stitch bonding 缝编法

35.hydroentangling [ˌhaɪdrəʊɪnˈtæŋglɪŋ]*n*. 水刺法

36.needle felt 针刺毡

37.adhesive binder 黏合剂

38.spraying [ˈspreɪɪŋ]*n*. 喷射；喷雾

39.foaming [ˈfəʊmɪŋ]*n*. 发泡

40.autogeneously [ɔːtəʊˈdʒiːnɪəsli]*adj*. 自发的；自生的

41.fuse or weld fibre 熔丝或熔接纤维

42.card and web lapper 梳理和铺网机

43.aerodynamically [ˌeərəʊdaɪˈnæmɪkli]*adv*. 空气动力学地

44.suspended [səˈspendɪd]*adj*. 悬浮的

译文

第 38 课　非织造布

1. 定义与分类

"非织造布"一词起源于半个多世纪前,非织造布往往被认为是传统纺织品的低价替代品,通常使用改造后的纺织机械梳理成网(干法)。非织造的短纤维加工过程省略了纺纱阶段,通过化学、机械或热的方法对纤维网进行黏合(固结)取代传统纺织品中纱线的机织(或针织)。

因此,现在我们熟知的非织造布行业是从纺织、造纸和聚合物加工行业发展起来的。非织造布行业不愿与传统纺织行业及其协会联系在一起,也不希望其产品被称为"非纸品"或"非塑品"。

这种用词不当造成的错觉让一些人认为非织造布是某种大宗商品,甚至是廉价的贸易商品,而事实往往恰恰相反。EDANA(欧洲一次性用品和非织造布协会)将非织造布定义为"由定向或随机取向的纤维制成的片材、网或棉絮"。为了区分湿法非织造布和湿法纸材料,做了如下区分,"纤维性含量重量的 50% 以上为纤维、纤维长径比高于 300"的材料为非织造布。北美非织造布工业协会 INDA 将非织造布描述为通过各种机械、热和 / 或化学过程将纤维或长丝缠结在一起的片状或网状结构。非织造布是一种工程织物,可作为一次性或短期使用产品,也可依照用途作为耐用、长期使用的产品。非织造布具有类似常规机织和纺织织物的外观、质地和强度。除了单层非织造布外,多层非织造复合材料、层压材料和三维非织造织物也已在商业化生产中。INDA 列出的最常见的非织造布产品包括:

(1)一次性尿布;

(2)卫生巾和卫生棉条;

(3)医疗领域使用的无菌包布、帽子、罩衣、口罩和窗帘;

(4)家用和个人湿巾;

(5)服装衬布;

(6)地毯和室内装潢织物、衬垫和背衬;

(7)农业覆盖物和种子条;

(8)汽车顶篷和内饰;

（9）土木工程织物／土工织物。

2. 原材料

在非织造布生产中，人造纤维占主导地位，占总产量的 90% 以上。人造纤维分为三类，即由天然聚合物制成的、由合成聚合物制成的和由无机材料制成的。全世界用于非织造布的纤维中，聚丙烯纤维为 63%，聚酯纤维为 23%，黏胶纤维为 8%，聚丙烯腈纤维为 2%，聚酰胺纤维 1.5%，其他特种纤维为 3%。聚丙烯纤维在非织造布工业中占主导地位。因水刺擦拭巾市场膨胀，人们认为黏胶纤维的使用份额会有所增加。在非织造布行业中，溶剂法纺制的纤维素纤维莱赛尔也变得重要了，部分原因在于其良好吸湿性和高湿强度。

3. 成网

在所有非织造成网过程中，纤维或长丝要么沉积在成型表面上成网，要么凝缩成纤网喂入输送机表面。这个阶段可以是干的、湿的或熔融的——干法、湿法或聚合物成网（也称为纺丝和纺熔法工艺）。成网涉及短纤维或长丝转化为二维（网）或三维网集合体（棉絮）过程，这是织物的前道。它的结构和成分极大地影响着成品织物的尺寸、结构和性能。纤维网中的纤维取向由纺织、造纸或聚合物挤出机器进行控制。纤维在纤维网中的排列（特别是纤维取向）决定了织物性能的取向性，大多数非织造布是各向异性的。尽管可直接测量纤网中的纤维取向，但通常是测量纤网或织物的纵向／横向（MD：CD）比率。

这种纵向（MD）和横向（CD）的织物性能（通常是拉伸强度）之比可反映织物中纤维的取向。实际生产中，很难获得各向同性结构（即 MD：CD=1）的网或织物，在技术上是非必要的。在成网阶段，其他关键织物参数也影响着半成品重量和加工宽度。传统上，每个成网系统针对特定的纤维或产品，但不同成网系统加工相近商业产品也越来越频繁。以高蓬松度无纺布为例，它可使用梳理机和交叉铺网机或辊式气流成网系统生产。在卫生用品方面，人们更喜欢使用梳理法和水刺法生产的柔软短纤维产品，以替代气流成网和热黏合产品。

4. 网固结

非织造布黏合工艺可使用机械的、化学的(包括使用溶剂的黏合)或热的。黏合度是决定织物的机械性能(特别是强度)、孔隙率、柔韧性、柔软度和密度(蓬松度、厚度)的主要因素。黏合可以单独工序进行,但通常是在成网过程中同步完成。在一些织物中,使用了多种黏合工艺。机械固结方法包括针刺法、缝编法和水刺法。在针刺方面,最常用喂入机械为梳理机和交叉铺网机,预计全球的针刺毡产量达到 110 万吨,其中超过 72% 针刺非织造布使用新纤维,而不是回收纤维。化学黏合方法是通过喷涂、印刷或发泡技术将黏合剂施加到网上。溶剂黏合是用适当的化学品软化或部分溶解纤维表面,以在交叉点进行黏合纤维或纤维自黏合。热黏合是使用热和压力来软化,将纤维熔合或熔接在一起,而非熔化。

5. 干法非织造

第一个干法系统起源于中世纪时期的毡制工艺。在压毡行业中,梳理和铺网机加工羊毛或羊毛混合絮垫,再加湿、搅动和加热进行毡化。非织造布行业中使用的一些干法成网技术,特别源于纺织业加工纤维的方法——梳理。在干法成网中,纤维被梳理(包括梳理和交叉铺网)或空气动力学成型(气流成网),然后由机械、化学或热方法黏合。这些方法是针刺、水刺、缝编(机械)、热黏合和化学黏合。

6. 湿法非织造

通过控制液体中悬浮的短纤维,用机械制造的类似纸的无纺布,被称为"湿法成网"。少数公司使用湿法成网工艺,其运行资金量大且耗水量大。除了纤维素纸,该法还可生产芳纶、玻璃和陶瓷等高性能纤维的技术纸。

Lesson 39 Three-dimensional fibrous assemblies

Textile structures such as in woven, knitted, nonwoven and braided fabrics are being widely used in advanced structures in the aerospace, automobile, geotechnical and marine industries. In addition, they are finding wide application as medical implants such as scaffolds, artificial arteries, nerve conduits, heart valves, bones, sutures, etc. This is because they possess outstanding physical, thermal and favourable mechanical properties, particularly light weight, high stiffness and strength, good fatigue resistance, excellent corrosion resistance and dimensional stability.

Three-dimensional woven fabrics are produced principally by the multiple warp weaving method, which has long been used for the manufacture of double and triple cloths for bags, waddings and carpets. The 3-D solid angle interlock principle involves the binding of straight warp yarns by interlocking warp yarns. Warp yarns can be bound to different depth. As in orthogonal fabrics, wadding yarns may be used in the structure. The structures of various 3-D solid woven fabrics are presented in Fig. 39.1.

In addition, 3-D fabrics act as attractive reinforcing materials in various composite applications with low fabrication cost and easy handling. With high-end applications such as in aerospace, the orientation of the fibrous reinforcement is becoming more and more important from a load-bearing point of view, as is the need for placing the reinforcement oriented in the third dimension. Textile fabrics, termed preforms in composites and other applications, consist of various reinforcing fabrics such as wovens, knits, braids and nonwovens.

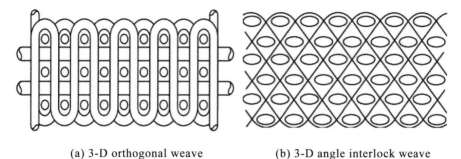

(a) 3-D orthogonal weave   (b) 3-D angle interlock weave

Figure 39.1   3-D solid woven structures

To extend the use and value of textiles into industrial and engineering applications, which typically require strength in more than two directions, textile designers have bound together layers of textiles and exploited the chemical properties of fibres and binders to create novel nonwoven textiles whose fibres are not restricted to two-dimensional arrangements. 3-D fabrics have been introduced to respond to the needs of a number of industrial requirements such as composites capable of withstanding multidirectional stresses.

The development of 3-D textiles has taken place rapidly over the past two decades. It can be credited largely to the growth of another technology: composite materials, which combine fibres and a matrix. Textile engineers have been challenged to develop strong fibre architectures and new manufacturing processes for building textile structures in three dimensions, as these 3-D fabrics hold great promise for use in industry, construction, transportation and even military and space applications. An understanding of the production methods and structures of these 3-D fibrous assemblies would go a long way in the design, process control, process optimization, quality control, clothing fabrication and the development of new techniques for specific end uses. The interrelationship between their structure and various properties may be of great help in designing new types of 3-D structures for the construction, medical, sports and aerospace Industries.

生词与词组

1.aerospace ['eərəʊspeɪs]*n.* 航空航天工业

2.geotechnical [ˌdʒiːəʊ'teknɪkəl]*n.* 岩土工程

3.marine [məˌriːn]*n.* 海洋业

4.implant ['ɪmplænt]*n.* 移植片，植入物

5.scaffold ['skæfəʊld]*n.* 支架

6.artificial artery 人造动脉

7.nerve conduit 神经导管

8.heart valve 心脏瓣膜

9.bone [bəʊn]*n.* 骨骼

10.suture ['suːtʃə(r)] *n.* 缝合线

11.fatigue resistance 抗疲劳性

12.corrosion resistance 耐腐蚀性

13.dimensional stability 尺寸稳定性

14.multiple warp 多经

15.double and triple cloths for bag 双层和三层布的包袋

16.angle interlock 角交联

17.orthogonal [ɔː'θɒgənəl]*adj.* 正交的；直角的

18.wadding yarn 填充纱

19.multidirectional [mʌltɪdɪ'rekʃənl]*adj.* 多方向的

20.architecture ['ɑːkɪtektʃə(r)]*n.* 体系结构

译文

第 39 课　3D 纺织品

　　机织、针织、非织造和编织织物等纺织结构材料正被广泛用于航空航天、汽车、岩土工程和海洋工业。此外，它们也广泛应用于医疗植入物，如支架、人工动脉、神经导管、心脏瓣膜、骨骼、缝合线等。这归因于它们所展现的优异物理、热学以及良好的机械性能，特别是重量轻、刚度和强度高、良好的抗疲劳性、优异的耐腐蚀性和尺寸稳定性。
　　三维机织物主要采用多经编织方法生产，该法长期用于制造双层和

三层布的包袋、填充材料和地毯。3D 立体交联是由交联经纱对直经纱的捆扎。经纱可以绑定到不同的深度。在正交织物中可使用填充纱。各种 3D 机织物的结构如图 39.1 所示。

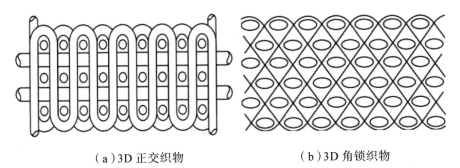

（a）3D 正交织物　　　　　　　　（b）3D 角锁织物

图 39.1　3D 多层织物结构

此外,3D 织物以优良增强材料应用于各种复合材料中,其制造成本低且易于处理。在航空航天等高端应用中,定向纤维增强材料在承载方面变得越来越重要,因为需要将增强材料定向放在三维空间中。在复合材料和其他应用中,纺织织物称为预制件,其由各种增强织物组成,如机织、针织、编织和无纺布。

为了拓展产业用纺织品(产业用纺织品需要两个以上方向的强度)的用途和价值,纺织品设计师将多层纺织品黏合在一起,并利用纤维和黏合剂的化学特性来创造新型非织造纺织品,其纤维不限于二维排列。为了满足许多工业要求,3D 织物正推广开来,如可承受多向应力的复合材料。

在过去的二十年中,3D 纺织品快速发展。这归功于复合材料技术的发展,复合材料是纤维和基质结合体。在工业、建筑、运输、军事和太空领域中,3D 织物具有广阔的应用前景,纺织工程师正尝试开发具有强纤维结构和三维纺织结构的制造工艺。了解这些 3D 纤维集合体的生产方法和结构,将有助于其设计、过程控制、过程优化、质量控制、服装制造及特定用途的技术开发。了解其结构和各种特性之间的关系,对建筑、医疗、体育和航空航天工业领域新型 3D 结构的设计有很大帮助。

Lesson 40 Dyeing

The colouration of textiles is a mature and highly efficient industrial technology. The consumption of dyes worldwide is some 360,000 tonnes per year comprising a major part of the $6 billion dollar dye industry. Unlike other textile fibres, there is very little colouration of fibre prior to spinning and the great majority of products are dyed and printed in fabric form. A number of distinct dyeing processes and classes of dye have been developed and are particularly suited to certain product types.

Principles of dyeing

The objective of dyeing is to produce uniform colouration of a substrate usually to match a pre-selected colour. The colour should be uniform through out the substrate and be of a solid shade with no unlevelness or change in shade over the whole substrate. There are many factors that will influence the appearance of the final shade, including: texture of the substrate, construction of the substrate (both chemical and physical), pre-treatments applied to the substrate prior to dyeing and post-treatments applied after the dyeing process. The application of colour can be achieved by a number of methods, but the most common three methods are exhaust dyeing, continuous (padding) and printing.

生词与词组

1.colouration [ˌkʌləˈreɪʃ(ə)n]*n.* 着色；染色
2.pre-selected colour 预选颜色
3.pre-treatment 前处理
4.post-treatment 后处理
5.exhaust dyeing 浸染
6.continuous (padding) 连续（轧染）

译文

第 40 课　染色

纺织品染色是一项成熟、高效的工业技术。全世界的染料消费量约为 360000 吨／年,其中价值 60 亿美元染料行业是其主要组成。不像其他纺织纤维,纺纱前纤维几乎无色,而多数纺织品均以织物形式进行染色或印花。现已开发了各种染色工艺及染料,且其特别适用于特定产品类型。

染色原理

染色的目的是按预定颜色使基材均匀着色。整个材料的着色是均匀且是纯色的,无不均匀或色调变化。有许多因素会影响最终的色调,这包括材料的质地、材料的结构(化学和物理)、染色前基材的预处理以及染色后基材的后处理过程。颜色的施用可由多种方法来实现,但最常见的三种方法是浸染、连续(轧染)和印花。

Lesson 41   Dyes

1.Direct dyes

In 1884 Bottiger discovered that the diazo dye, Congo Red, coloured without the necessity for pre-treatment with a metal salt (a so-called "mordant"). This finding led to the synthesis of related dyes which were referred to as the "direct" dyes due to their ease of application. The dyes are generally sulfonated poly-azo compounds although other structures such as metal complexes and anthraquinones are utilised to complete the shade palette. The levels of wash-fastness achieved using direct dyes is generally not as high as the vat dyes, but their ease of application and broader palette led to this dye class being of great importance until the discovery of the reactive dyes. The direct dyes generally have better lightfastness than the corresponding reactive dyes and so find particular use in applications where laundering is infrequent but resistance to fading is desirable, e.g., curtains.

2.Reactive dyes

Reactive dyes are dyes that form a covalent bond with cotton fibres. The key parts of the dye molecule are the chromophore and the reactive group. ICI company released the first range of reactive dyes—the Procions in 1956, which were dyes based on the dichlorotriazinyl reactive group. These dyes were closely followed by Ciba's monochlorotriazinyl based Cibacron dyes. These dyes were enthusiastically embraced by cotton processors, and there followed an intense research and development effort which led to all the major textile dye manufacturers producing ranges of cotton reactive dyes.

3.Vat, sulphur and azoic dyes

Vat dyes derived from natural sources are the oldest dyes known. Synthetic vat dyes and modern versions of the vat dyeing process are highly important for the colouration of fibre. The application of sulphur dyes has similarities to the vat dyeing process and is particularly important for deep shades. Azoic dyes are insoluble dyes formed in situ by the same reactions used to synthesise azo dyes. The dyes have in common the property that they are applied by a two-step process in which water-soluble forms of the dye are absorbed by cotton and subsequently aftertreated to yield insoluble dyes in the fibre. These dyeing processes lead to dyed cotton goods with very high fastness and comprise a very significant part of the colouration industry.

4.Dye pollution

Environmentally responsible dye application involves the principles of pollution prevention that were developed and promulgated in the early 1990s with the hierarchy of "reduce, reuse, recycle". This replaced the earlier "end-of-pipe" response to growing environmental legislation. A common mantra of the environmentally concerned is to "think globally, act locally", and this readily applies to dyeing and associated operations. A dyehouse may have limited impact on the immediate locality if its air emissions and wastewater are uncontaminated (or minimally so). For the former, volatile organic compounds (VOCs) and odour are the usual concerns. The main problems centre on water, where biological and chemical oxygen demand (BOD, COD), pH, total dissolved solids (TDS), temperature, oil/grease, heavy metals and colour are typically regulated.

Environmental acceptability in textile products generally falls into one of two categories. The first and simplest to demonstrate is that the product will not harm the user, or harm the environment in use. A primary example is the Oeko-tex 100 scheme that certifies items being sold as environmentally sound, based on what is present or might be released from them. The second category of greenness is based on the environmental impact of the production of the item: from cotton field or fibre factory to end use, and beyond to ultimate disposal.

生词与词汇

1.diazo dye 偶氮染料

2.mordant [ˈmɔ:dnt]n. 媒染剂

3.sulfonated poly-azo 磺化多偶氮

4.metal complex 金属络合物

5.anthraquinone [ˌænθrəkwɪˈnəʊn]n. 蒽醌

6.shade palette 暗沉色调

7.wash-fastness 水洗牢度

8.vat dye 还原染料

9.broader palette 色调多

10.lightfastness 耐日晒牢度

11.covalent bond 共价键

12.chromophore 发色团

13.reactive group 反应基团

14.dichlorotriazinyl reactive group 二氯三嗪基反应基团

15.monochlorotriazinyl 一氯三嗪

16.Cibacron dye 汽巴克隆染料

17.enthusiastically embraced 受热烈欢迎的

18.intense research 深入研究

19.sulphur dye 硫化染料

20.deep shade 深色调

21.water-soluble form 水溶性形式

22.insoluble dye 不溶性染料

23.promulgate ['prɒmlgeɪt]*v.* 发布；发表

24.hierarchy ['haɪərɑːkɪ]*n.* 制度；层次体系

25.mantra ['mæntrə]*n.* 口头禅

26.dyehouse ['daɪˌhaʊs]*n.* 印染厂

27.volatile organic compound 挥发性有机化合物

28.odour ['əʊdə(r)] *n.* 气味

29.total dissolved solid 总溶解固体

30.disposal [dɪ'spəʊzl]*n.* 处理；处置

31.Congo Red 刚果红

译文

第 41 课　染料

1. 直接染料

1884 年,保蒂格(Bottiger)发现偶氮染料——刚果红不需用金属盐(所谓的 "媒染剂")进行前处理即可直接染色。此发现促进了相关染料的合成,这些染料可直接应用而被称为 "直接" 染料。直接染料多为磺化的多偶氮化合物,其他结构如金属配合物和蒽醌用于显色。直接染料的水洗牢度等级通常不及还原染料,但它们使用方便且色调多,在活性染料发现前,该染料占有重要地位。直接染料往往比相应的活性染料具有更好的耐日晒牢度,因此其特别适用于不经常洗涤且需抗褪色需求,如窗帘。

2. 活性染料

活性染料是一种可与棉纤维形成共价键的染料。染料分子的关键部分是发色团和活性基团。1956 年, ICI 公司发布了第一批二氯三嗪基活性染料——Procions。紧随其后的是汽巴公司的一氯三嗪基汽巴克隆染料。这些染料受到棉花加工商的热烈欢迎。随后大部分染料制造商进行了大量的研究和开发,并生产了一系列棉用活性染料。

3. 还原染料、硫化染料和偶氮染料

源自自然的还原染料是当前最古老的染料。合成还原染料及其当前的还原染色工艺对纤维的着色非常重要。硫化染料的应用与还原染色工艺相似,尤其适用于染深色调。偶氮染料由相同反应原位制成合成偶氮染料,其为不溶性染料。这些染料的共同特性是染色需两步完成,形成水溶性染料并被棉纤维吸收,再通过后处理形成不溶性染料固着在纤维上。该染色工艺赋予了棉织物良好的色牢度,是染色工业的重要组成部分。

4. 染料污染

环境友好的染料应用涉及 20 世纪 90 年代初制定和颁布的"减少、再利用、回收"层次体系的污染预防原则。这取代了先前的"管道末端"的环境立法。一句环保格言是"放眼全球,本地行动",其很适用于染色和相关操作。如果染厂的废气和废水未污染(或最低限度)周边地区,当下其对周边地区的影响有限。就废气而言,挥发性有机化合物(VOCs)和气味是常见的问题。主要问题是废水,其带有的生物和化学需氧量(BOD、COD)、pH、总溶解固体(TDS)、温度、油/油脂、重金属和颜色会被监管。环境友好的纺织产品一般分为两类。第一类为该产品应被证实其对使用者或环境无损害。Oeko-tex 100 计划是典型案例,该计划基于物品中留存或可能释放的物质来确保出售物品的环保可靠性。第二类为基于环境影响的绿色生产产品:从棉田或纤维工厂到最终使用,再到最终处置。

Lesson 42　Exhaust dyeing

In exhaust dyeing, the dye, which is wholly or partially soluble in the dye bath, is transported to the fibre surface by the motion of the dye liquor or by the motion of the substrate being dyed. The dye is adsorbed onto the fibre surface and ideally diffuses into the whole of the fibre. Depending upon the dye being used, the interactions between the dye and the fibre can be either chemical or physical. Exhaust dyeing is usually conducted using dilute solutions of dyes, normally termed long liquor dyeing, and can involve liquor to substrate ratios from 8:1 up to 30:1. As described above there are two main phases to exhaust dyeing, the adsorption phase and the diffusion phase. Most exhaust dyeing involves a temperature gradient whereby the dyeing is commenced at a fairly ambient temperature (30 ℃ –40 ℃) with the temperature being increased slowly up to a final temperature which is dependent upon the dyes being used. Depending upon the dyes being used, during the diffusion phase, changes to the dye bath pH may be made to facilitate covalent fixation of the dye which has diffused into the substrate. Typical, but not mandatory, aspects of modern exhaust dyeing equipment are pumped circulation of the dye liquor, a sealed system which can be pressurised, microprocessor control of heating and flow.

An example of modern exhaust dyeing machinery is the package dyeing machine shown schematically in Fig. 42.1. Yarn wound onto perforated centres is loaded onto a central shaft and compressed to a uniform density. The liquor is circulated via a uniform radial flow from the perforated shaft, through the yarn package into the main part of the vessel and recirculated via the pump; this flow direction can be reversed and is usually alternated during the course of a dyeing to ensure uniform contact between the yarn and the dye liquor. The other major types of exhaust dyeing machinery are those used for dyeing cotton in fabric form. There are four main types, the beam, beck, jet and jig. Fabric is wound onto a perforated shaft and the dye liquor is pumped radially through the fabric roll in a similar manner to the package dyeing machine discussed above.

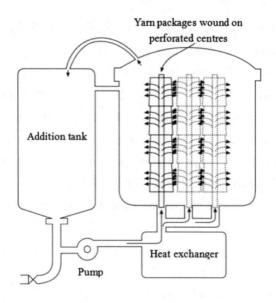

Figure 42.1　Package dyeing machine

生词与词组

1.dye bath 染浴

2.dye liquor 染液

3.dilute solution 稀溶液

4.adsorption phase 吸附阶段

5.diffusion phase 扩散阶段

6.gradient [ˈɡreɪdiənt]*n.* 坡度；梯度

7.covalent fixation 共价结合

8.mandatory [ˈmændətəri]*adj.* 强制的

9.sealed system 密封系统

10.pressurised [ˈpreʃəraɪzd]*adj.* 加压的；受压的

11.microprocessor control 微处理器控制

12.perforated [ˈpɜːfəreɪtɪd]*adj.* 带孔的；穿孔的

13.shaft [ʃæft]*n.* 柄；轴

14.radial flow 径向流

15.perforated shaft 孔轴

16.recirculate [rɪˈsɜːkjʊleɪt]*v.* 再循环

17.pump [pʌmp]*n.* 泵

18.beck [bek]*n.* 绳状

19.jet [dʒet]*n.* 射流

20.jig [dʒɪg]*n.* 卷染

译文

第 42 课   浸染

在浸染染色中,染料完全或部分溶解在染浴中,染液或被染物的运动使染料扩散至纤维表面。染料吸附在纤维表面,并扩散至整根纤维。按照使用的染料,染料和纤维之间的相互作用可以是化学的,也可以是物理的。浸染通常使用稀释染液进行,常称为大浴比染色,其浴比为 8∶1 到 30∶1。如上所述,浸染有两个主要阶段,吸附阶段和扩散阶段。多数浸染涉及温度梯度,从室温(30℃ –40℃)开始,温度缓慢升高至最终温度,最终温度由使用的染料决定。根据使用的染料,在扩散阶段,改变染浴 pH 值可促进已扩散到材料上的染料与材料共价结合。典型而非强制,现代浸染设备设有染液的泵循环输送的密封系统,系统内压力和流量、加热由微处理器控制。

现代浸染机械筒纱染色机,如图 42.1。纱线缠绕在带孔的筒管上,安装到中心轴并堆叠一致。染液由多孔轴沿径向均匀分散,穿过筒纱进入容器内,再经泵循环;这种流向可颠倒,在染色过程中通常可交替进行,以确保纱线和染液之间的均匀接触。其他类型的浸染机械是用于对棉织物染色的机械。有四种主要类型:经轴染色、绳状染色、射流染色和卷染。织物被缠绕在孔轴上,以上述卷装染色机类似的方式染液由泵压送沿径向穿过织物辊。

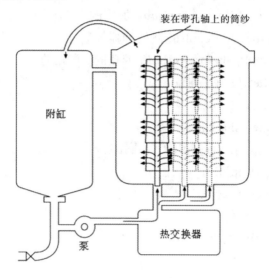

图 42.1　筒纱染色机

Lesson 43   Continuous dyeing

1.Semi-continuous dyeing

In the early 1960s the cold pad-batch process was developed for the application of reactive dyes to cotton and has since found extensive adoption in the dyeing of cotton fabrics. The cold pad-batch process takes advantage of the ability of many reactive dyes to react with cotton at room temperature in practical time scales. In this method the fabric is impregnated ("padded") with a dye paste and stored ("batched") for up to 24 hours at ambient temperatures to allow fixation to take place. The dyed fabric is then washed off.

2.Continuous dyeing

Continuous dyeing is a method of dyeing fabrics in which, in an uninterrupted sequence, they are first impregnated with dyes and chemicals followed by a fixation step and rinsing and drying. The impregnation of fabric with dye is generally carried out in a padder. Fixation can occur by a number of mechanisms such as steaming, baking or simply exposure to the atmosphere. Steaming is a very common component of continuous dyeing ranges for fabrics, the aqueous high-temperature environment allowing the diffusion of dyestuff molecules into the fibre.

Continuous dyeing is a process whereby dyeing the fabric and fixation of the dye are carried out continuously in one simultaneous operation. This is traditionally accomplished using a production line system where units are assembled into lines of consecutive processing steps; this can include both pre- and post-dyeing treatments. Fabric is usually processed in open width, so care must be taken not to stretch the fabric. The fabric running speed dictates the dwell time of the fabric through each treatment unit, although dwell times can be increased by using "festoon" type fabric transport. The main disadvantage to continuous processing is that any machinery breakdown can cause ruined fabric due to excessive dwell times in specific units whilst the breakdown is being rectified; this can be a particular problem when stenters running at high temperatures are employed since fabrics may be severely discoloured or burnt.

3.Air dye technology

Air dye technology adds colour to textiles without using water. The process of making textiles can require several dozen gallons of water for each pound of clothing. The air dye process employs air instead of water to help dyes penetrate, a process that uses no water and requires less energy than traditional methods of dyeing. This technology works only on synthetic materials, not on natural fibre products. Advantages of this technology are that it is reliable and does not pollute water in the colour application process. No hazardous waste is emitted while replacing water by air and no water is wasted. The management of air dye technology greatly reduces energy requirements, thereby lowering costs and satisfying the strictest standards of global responsibility. There is no need to engage boilers, screen printing machines, drying ovens, or cleaning and scouring chemicals, which eliminates major sources of pollution. Eliminating water in the colour application step simplifies the process, creating revolutionary possibilities of new industry and employment in unfarmable, arid regions of the world.

生词与词组

1.pad-batch 浸轧

2.impregnate [ɪmˈpregneɪt]v. 浸泡

3.ambient [ˈæmbiənt]n. 环境　adj. 外界的；周围的

4.fixation [fɪkˈseɪʃ(ə)n]n. 固色；固着；固定

5.rinsing [ˈrɪnsɪŋ]n. 漂洗；冲洗；漂掉

6.padder [ˈpædə]n. 轧车；浸轧机

7.dyestuff [ˈdaɪˌstʌf]n. 染料

8.simultaneous operation 同时操作；同步操作

9.pre- and post-dyeing treatments 染色前和染色后处理

10.dwell time 停留时间

11.festoon [feˈstuːn]n. 花彩；花饰

12.breakdown [ˈbreɪkdaʊn]n. 崩溃；故障；死机

13.rectified [ˈrektɪfaɪd]v. 修复；改正

14.stenter [ˈstentə]n. 拉幅机

15.discoloured [dɪsˈkʌlə(r)d]adj. 已变色的；已褪色的

16.engage boiler 锅炉

17.screen printing machine 筛网印花机

18.drying oven 烘箱；干燥炉

19.scouring chemical 精炼化学品

译文

第43课　连续染色

1. 半连续染色

20世纪60年代初期，冷轧堆工艺用于活性染料染棉，此后在棉织物染色中得到广泛采用。利用许多活性染料于棉纤维的反应能力，冷压堆工艺在室温及一定时间下进行。在这种方法中，织物浸泡（"浸轧"）在染液中，并在室温下储存（"成批"）24小时以进行固色，然后水洗。

2. 连续染色

连续染色是一种不间断的织物染色工艺,织物浸渍染料和化学品,再进行固色、漂洗和干燥。织物浸渍染料通常在轧车中进行。可通过多种机制进行固色,如蒸汽固色、烘烤固色或暴露在空气中。蒸汽固色是织物连续染色中较为常见的部分,高温蒸汽环境下染料分子易于扩散至纤维中。

连续染色是将织物染色、固色在一个同步操作中连续进行的过程。传统连续染色在连续生产线中完成,各单元按照加工流程进行集成,这也包括染前和染后处理。织物常以平幅加工,应注意不要拉伸织物。织物运行速度决定了织物通过每个处理单元的停留时间,"花"型织物传输需增加停留时间。连续加工的缺点是任何机器故障均可能导致织物损坏,在修复故障时特定单元的停留时间过长;特别是在使用高温下运行的拉幅机时,织物可能会严重变色或燃烧。

3. 空气染色技术

空气染色技术无须用水即可为纺织品增添色彩。加工一磅衣服需使用几十加仑的水。空气染色工艺使用空气而不是水携带染料渗透,这种工艺不使用水且比传统染色方法更节能。该技术仅适用于合成材料,不适用于天然纤维产品。该技术的优点是可靠,上色过程不产生废水。空气代替水不产生危险废物,不浪费水。使用空气染色技术可大大降低能源需求,从而降低成本,并达到最严格的环保标准。不需使用锅炉、筛网印花机、烘箱、清洗和精炼化学品,从而消除了主要污染源。该技术上色阶段消除水,简化了过程,为世界上不可耕种的干旱地区创造了革命性的新工业和就业。

Lesson 44　Printing of fabrics

Printing has always been an important process for producing fashion effects on fabrics. The technology of printing has changed over the years and is currently experiencing a period of rapid development. The use of carved wooden blocks for printing fabrics has been known since antiquity. The technique was used on a craft or semi-industrial basis and techniques were highly varied and regional. Multi-step processes included the printing of metal salts (mordants) or alternatively waxes and other "dye resist" compounds. Subsequent dyeing yielded a print in which the dye fixed to mordanted regions and areas printed with dye resist remained undyed. The technique of applying dyes onto fabrics to obtain printed fabrics directly was a relatively late development.

There are a wide number of methods for applying dyes by printing techniques. The most common method is direct printing whereby the dyes are applied in the form of a print paste containing thickeners and auxiliaries. The print paste is applied to the fabric via a roller, in the case of engraved cylinders, or by screens in the case of flat or rotary screen printing. Following printing the prints arc dried and steamed similar to the processes used for padding, depending upon the dyes being printed. Each colour of the design requires its own screen, so printing machines can take up a significant amount of space, especially a flat screen machine. Pigment printing involves the use of binders to adhere the pigment to the substrate surface. Pigment prints usually only require drying/curing after the printing process. Ink-jet printing is rapidly becoming a popular method of producing patterned designs on fabrics. Production speeds are not as fast as rotary screen printing, with machines running at 20m/min–30m/min depending on the complexity of the design being printed. Ink-jet printing poses no problems with soluble dyes for printing substrates such as cotton, nylon, wool, etc., but disperse dyes and pigments must be finely dispersed within the printing ink to avoid blocking of the ink-jet nozzles which can be costly to replace and time consuming to unblock.

生词与词组

1.fashion effect 时尚效果
2.semi-industrial basis 半工业基础
3.dye resist 防染色
4.mordanted region 媒染区域
5.undyed [ʌnˈdaɪd]*adj.* 未染色的
6.thickener [ˈθɪkənə(r)]*n.* 增稠剂；稠化剂；增厚剂
7.engraved cylinder 压花辊筒；印花辊筒；沟槽辊筒
8.rotary screen printing 旋转筛网印花；圆网印花
9.pigment printing 颜料印花
10.ink-jet printing 喷墨打印
11.disperse dye 分散染料
12.ink-jet nozzle 喷墨喷嘴

译文

第44课　织物印花

印花一直以来是在面料上产生时尚效果的重要工艺。多年来，印花技术已发生质变，正快速发展。自古以来，人们就使用雕刻木块来印制织物。该技术是在手工艺或半工业基础上使用的，具有多样性、区域性。多步工艺包括金属盐（媒染剂）或蜡和其他"防染色"化合物印花。染色后印花，在媒染区域固着染料，印有防染剂的区域保持未染色。使用染料在织物上直接印花的技术发展相对较晚。

现有多种染料印花的技术方法。最常见的方法是直接印花,其由染料与增稠剂、助剂形成印花浆料使用。印花浆料由印花辊筒(雕刻花纹的滚筒)渗出至织物上,或者在平网、圆网下进行筛网印花。印花后,印花品依据染料类别需进行干燥和汽蒸,类似用于轧染的过程。每种颜色的设计都需单独的筛网,因此印花机会占用大量空间,特别是平网印花机。颜料印花涉及使用黏合剂将颜料黏附到材料表面。颜料印花通常只需在印花后进行干燥/固化。喷墨印花正迅速成为织物图案设计的流行方法。喷墨印花生产速度不如圆网印花快,机器的运行速度为20-30米/分,这取决于印花图案的复杂程度。喷墨印花适用于可溶性染料印制棉、尼龙、羊毛等材质,但分散染料和颜料需分散在印花油墨中,以防止堵塞喷墨喷嘴,喷嘴更换成本高昂,清除堵塞时间长。

Lesson 45　Chemical finishing

1.Wet and dry or chemical and mechanical finishing

Textile wet processing can be thought of having three stages, pre-treatment (or preparation), colouration (dyeing or printing) and finishing. Finishing in the narrow sense is the final step in the fabric manufacturing process, the last chance to provide the properties that customers will value. Finishing completes the fabric's performance and gives it special functional properties including the final "touch". But the term finishing is also used in its broad sense: "Any operation for improving the appearance or usefulness of a fabric after it leaves the loom or knitting machine can be considered a finishing step." This broad definition includes pre-treatments such as washing, bleaching and colouration. The term finishing is used in the narrow definition to include all those processes that usually follow colouration and that add useful qualities to the fabric, ranging from interesting appearance and fashion aspects to high performance properties for industrial needs. This definition may be applied to similar finishing processes for grey fabrics (without colouration). Bleaching and carbonisation are chemical treatments that also improve the quality of fabrics.

Most finishes are applied to fabrics such as wovens, knitwear or nonwovens. But there are also other finishing processes, such as yarn finishing, for example sewing yarn with silicones and garment finishing. Textile finishing can be subdivided into two distinctly different areas, chemical finishing and mechanical finishing. Chemical finishing or "wet finishing" involves the addition of chemicals to textiles to achieve a desired result. Physical properties such as dimensional stability and chemical properties such as flame retardancy can both be improved with chemical finishing. Typically, the appearance of the textile is unchanged after chemical finishing. Mechanical finishing or "dry finishing" uses mainly physical (especially mechanical) means to change fabric properties and usually alters the fabric appearance as well. Mechanical finishing also encompasses thermal processes such as heat setting. Typical mechanical finishes include calendering, emerising, compressive shrinkage, raising, brushing and shearing, and especially for wool fabrics milling, pressing and setting with crabbing and decatering.

Often mechanical and chemical finishing overlap. Some mechanical finishes need chemicals, for example milling agents for the fulling process or reductive and fixation agents for the decatering of wool fabrics. On the other hand chemical finishing is impossible without mechanical assistance, such as fabric transport and product application. The assignment to mechanical or chemical finishing depends on the circumstance, whether the major component of the fabric's improvement step is more mechanical- or chemical-based.

2.The challenge and charm of chemical finishing

The proper formulation of chemical finishes requires consideration of several important factors: the type of textile being treated (fibre and construction); the performance requirements of the finish (extent of effect and durability); the cost to benefit ratio; restrictions imposed on the process by availability of machinery, procedure requirements, environmental considerations; and compatibility of different formula components as well as the interaction of the finishing effects. To bring all these parameters to an acceptable compromise is not easy, even for a single purpose finish. But usually several types of finishes are combined for economical reasons mostly in one bath (only one application and drying process). This is often the hardest challenge of chemical finishing. First, all components of the finish bath must be compatible. Precipitations of anionic with cationic products have to be avoided. The emulsion stability of different products may be reduced by product interactions. More difficult is often the second hurdle, the compatibility of the primary and secondary effects of the different types of finishes that are being combined:

• Some effects are similar or assist each other, for example silicone elastomers cause water repellency, and softeners bring about antistatic effects and antistatic finishes can be softening.

• Some effects are obviously contradictory, for example hydrophobic finishes and hydrophilic antistatic finishes, or stiffening and elastomeric finishes, or stiffening and softening finishes.

• Other types of finishes typically reduce the main effect of a finish type, for example the flame retardant effect is decreased by nearly all other types of chemical finishes as they add flammable components to the fabric.

• Fortunately true antagonistic effects are rare, but true synergistic effects are also rare, where the resulting effect of a combination is greater than the sum of the single effects of the combined products. Examples of both cases are different types of flame retardants.

3.Importance of chemical finishing

Chemical finishing has always been an important component of textile processing, but in recent years the trend to "high tech" products has increased the interest and use of chemical finishes. As the use of high performance textiles has grown, the need for chemical finishes to provide the fabric properties required in these special applications has grown accordingly.

生词与词组

1.grey fabric 坯布

2.silicone [ˈsɪlɪkəʊn] *n.* 有机硅

3.flame retardancy 阻燃性

4.calendering [kæˈlɪndərɪŋ]*n.* 轧光；压花

5.emerising [ˈemˌraɪzɪŋ]*adj.* 仿麂皮的

6.compressive shrinkage 机械预缩

7.raising [ˈreɪzɪŋ]*n.* 起绒（毛）

8.brushing [ˈbrʌʃɪŋ]*n.* 刷绒（毛）

9.shearing [ˈʃɪrɪŋ]*n.* 剪毛

10.milling [ˈmɪlɪŋ]*n.* 缩呢

11.pressing [ˈpresɪŋ]*n.* 压呢

12.crabbing and decatering 煮呢和蒸呢

13.milling agent 缩绒剂

14.precipitation [prɪˌsɪpɪˈteɪʃ(ə)n]*n.* 沉淀

15.anionic [ˌænaɪˈɒnɪk]*adj.* 阴离子的

16.cationic [ˌkætaɪˈɒnɪk]*adj.* 阳离子的

17.emulsion stability 乳液稳定性

18.silicone elastomer 有机硅弹性体

19.water repellency 防水性

20.antistatic finish 抗静电整理

21.hydrophobic finish 疏水整理

22.hydrophilic antistatic finish 亲水抗静电整理

23.stiffening and elastomeric finish 硬挺和弹性整理

24.stiffening and softening finish 硬挺和柔软整理

25.flame retardant effect 阻燃效果

26.antagonistic effect 不兼容效果

27.synergistic effect 协同效应

译文

第45课　化学整理

1. 湿法、干法、化学、机械整理

纺织品湿法加工可分三个阶段：预处理（或准备）、着色（染色或印花）和整理。狭义的整理是织物加工的最后一步，是提供客户关注特性的最后机会。整理完善了织物的性能，并赋予其特殊的功能，包括最终的"触感"。术语整理的广义含义："在织物离开织机或针织机后，任何用于改善织物外观或实用性的操作均可以认为是整理步骤。"这个广义定义包括预处理，如洗涤、漂白和着色。术语整理的狭义定义包括所有在染色之后并为织物增加有用品质的过程，从有价值外观、时尚方面到满足工业需求的高性能特性。坯布（无染色）适用于该定义的类似整理工艺。漂白和碳化是化学处理，也可提升织物的质量。

多数整理应用于织物，如机织、针织或非织造布。但也有其他整理工艺，如纱线整理（如有机硅缝纫纱线）和服装整理。纺织品整理可分为两个截然不同的领域，化学整理和机械整理。化学整理或"湿整理"涉及向纺织品添加化学品以达到预期效果。物理性能（如尺寸稳定性）和化学性能（如阻燃性）都可由化学整理来改善。经化学整理后，纺织品的外观一般不会变化。机械整理或"干整理"主要使用物理（尤其是机械）方法来改变织物特性，通常也会改变织物外观。机械整理还包括热加工，如热定型。典型的机械整理包括轧光、仿麂皮整理、机械预缩整理、起绒（毛）、刷绒（毛）和剪毛，尤其适用于羊毛织物的缩呢、压呢、定型伴随煮呢和蒸呢。

通常机械和化学整理部分重叠。一些机械整理需要化学品，如用于缩绒过程的缩绒剂或用于羊毛织物蒸呢的还原剂和固定剂。另一方面，化学整理不可能没有机械辅助，如织物运输和产品应用。机械或化学整理的分配取决于实际情况，即织物改进步骤的主要部分更基于机械还是化学。

2. 化学整理的挑战与魅力

化学整理剂的恰当配方需考虑几个重要因素：被处理的纺织品类

型(纤维和结构),整理的性能要求(效果和耐久性的程度),成本效益,机器的可用性、程序要求、环保对过程施加的限制,不同配方成分的相容性以及整理效果的相互作用。即使出于单一目的,这些参数全都达到可接受也并不容易。但通常出于经济原因,通常将几种整理合并为一浴(仅一个施用和干燥过程),这通常是化学整理最困难的挑战。首先,整理浴的所有成分必须相容。必须避免产品中的阴离子与阳离子反应产生沉淀。不同产品的乳化稳定性会因产品相互作用而降低。更困难的往往是第二个障碍,不同类型整理被组合后,其主、次效果的兼容性:

• 一些效果相似或相互促进,如有机硅弹性体具有拒水性,柔软剂具有抗静电作用,抗静电整理具有软化作用。

• 一些效果明显矛盾,如疏水整理和亲水抗静电整理,或硬挺和弹性整理,或硬挺和柔软整理。

• 其他类型的整理剂通常会降低整理剂类型的主要效果,如几乎所有其他类型的化学整理剂都会降低阻燃效果,因为它们将易燃成分添加给织物。

• 幸运的是,不兼容作用较少,真正起协同作用的也较少,组合产生的效果大于组合产品的单一效果之和。

3. 化学整理的重要性

化学整理一直是纺织品加工的重要组成部分,但近年来"高科技"产品的发展趋势加强了对化学整理剂的青睐和使用。随着高性能纺织品需求增加,对提供特殊功能应用的化学整理剂的需求也相应增加。

Lesson 46   Carpet

Historically, the first floor coverings were probably animal skins and the earliest floor coverings of textile construction were probably crudely woven textiles made from rushes or grasses.

1.Classification of carpet fibres

Traditionally, carpets were produced using natural fibres until the advent of artificial and synthetic fibres, which had a significant impact on the carpet weaving industry. But new fibres have not played a very influential or critical role in the development of the industry, as far as tufted carpets are concerned. The most common carpet pile fibres which are currently in use are wool, silk, propylene and nylon and the common fibres used in backing are cotton, jute and polyester. Polypropylene fibre is the only one of its kind that is extensively used both on pile surface as well as backing.

2.Structure and properties of carpet fibres

A variety of natural as well as synthetic fibres are being used in textile floor coverings. Although there are many differences between a carpet fibre and an apparel fibre, the most obvious difference is the fibre diameter, since a higher diameter fibre is used in carpet. Apart from the diameter, there are quite a few parameters that play a decisive role in choosing a particular fibre for a specific product. This is applicable for both natural as well as synthetic fibres, e.g., cut length, luster, type of cross section, crimp, dyeability property; these are very important properties when selecting a nylon fibre. Similarly, the fibre micron, bulk, medullation, vegetable matter content, base colour, and fibre length after carding are the main parameters for selecting a wool fibre.

3.Carpet yarn engineering

Engineering of woollen yarns to produce the quality yarn to be used in carpet manufacturing is a tricky subject, and many aspects are to be taken into consideration while formulating a blend. For example, a buyer is concerned with price and yield, the dyer with colour and dyeability, the spinner with length and spinnability and the carpet manufacturer with yarn properties, carpet appearance and performance. Thus, selection of wool, blending, spinning and finishing play a critical role in trying to satisfy all these requirements. For convenience, yarn engineering can be divided into three parts: wool selection and blending, spinning and twisting, yarn finishing.

4.New trend

For a long time, high production output and high speed were the most important factors in the new developments for the carpet weaving industry. Carpet weavers want to weave different styles and change quickly from one quality to another. Flexibility is as important as productivity. Developments in raw materials contribute positively to this new trend. New dyeing techniques, chemical compounds, and treatments improve the quality of the yarns but also the choice of raw material. Another definite contributor is the new developments and technical improvements of the carpet-weaving machines. In the face-to-face weaving technique, there is a growing tendency towards carpets with a finer reed density and with more colours. Nowadays, it is already possible to weave carpets with 700 reed dents/m and eight colour frames. Serious efforts have been made to give a clearer backside to the carpets and to weave a carpet with a backside like a hand-knotted carpet and without warp yarns visible at the backside of the carpet. There is also a growing trend to produce carpets with relief, high-low effects, high pile or shaggy effects, combinations of cut pile, loop pile and flat weave effects, etc. Designers want to draw carpets using their full creativity without any limitations. A complete new revolution in Axminster weaving is the first high-speed Axminster weaving machine in 16 colour frames.

5.Residential use

Carpets for residential use are made throughout the world with particularly important centres in the USA and Western Europe. Important centres of carpet manufacture are emerging in the Indian sub-continent and in China. The products made in each of these centres have evolved in different ways and display different characteristics of style. In general terms, US carpets often have a pile of polyamide in patterned and textured loop pile, polyester in longer cut-pile referred to as "Saxony's" and more budget-conscious constructions with polypropylene pile.

6.Commercial use

Carpets intended for commercial use are often subjected to greater concentrations of traffic and need to withstand this. Carpets for offices, particularly modern open-plan offices and for some larger retail premises, feature hard-wearing dense low loop-pile constructions usually with polyamide pile. The carpet tile has found particular favour for this end use. Carpet tiles, frequently available in 50 square centimetres or 45.7 square centimetres (18 squre inches) formats, lend themselves particularly to large multi floor installations. The original carpet tiles were of a simple felt face layer bonded to the tile backing. Styles have developed from these through needled floor coverings, plain loop-pile tufted, plain cut-pile tufted and even patterned constructions, at each stage gaining in style, luxury and sophistication.

生词与词组

1.floor covering 地板覆盖物

2.crudely [ˈkruːdli]*adv.* 粗糙地

3.rush [ˈrʌʃ]*n.* 灯芯草

4.vegetable matter content 植物质含量

5.carpet-weaving machine 地毯织机

6.face-to-face weaving technique 面对面编织技术

7.fine reed density 高筘密

8.hand-knotted carpet 手工打结地毯

9.shaggy effect 蓬松效果

10.cut pile 割绒

11.loop pile 圈绒

12.sub-continent 次大陆

13.budget-conscious[ˈbʌdʒɪt ˈkɒnʃəs]*adj.* 更经济的；有预算的

14.polypropylene pile 聚丙烯绒

15.retail premise 零售场所

16.hard-wearing dense 紧密耐磨

17.loop-pile construction 小绒圈结构

18.carpet tile 地毯砖；地毯块

19.multi-floor installation 多层设施

20.needled floor covering 针刺地面覆盖物；针刺地毯

21.plain loop-pile tufted 平针圈簇绒

22.plain cut-pile tufted 平针割簇绒

23.sophistication [səˌfɪstɪˈkeɪʃ(ə)n]*n.* 精致；先进

译文

第46课 地毯

历史上,最早的地板覆盖物可能是动物皮,而最早的纺织结构地板覆盖物可能是由灯芯草或草制成的粗纺纺织品。

1. 地毯纤维的类别

传统地毯是使用天然纤维制造的,直到人造和合成纤维出现,这对地毯编织行业产生了重大影响。就植绒地毯而言,新纤维在行业内并未发挥非常有影响力或关键的作用。目前,最常见的地毯绒纤维是羊毛、丝绸、丙烯和尼龙,用于背衬的常见纤维是棉、黄麻和聚酯。聚丙烯纤维是唯一一种广泛用于绒表和背衬的纤维。

2. 地毯纤维的结构和性能

各种天然和合成纤维被用于铺地织物。地毯纤维和服装纤维之间存在很多差异,最明显的差异则是纤维直径,地毯纤维的直径更大。除了直径之外,特定产品需依据相应参数选择特定纤维。这既适用于天然纤维,也适用于合成纤维,如切割长度、光泽、横截面类型、卷曲、染色性能;选择尼龙纤维,这些是非常重要的特性。同样,纤维直径、体积、髓腔、植物质含量、基色和梳理后的纤维长度是选择羊毛纤维的主要参数。

3. 地毯纱工程

用于地毯制造的优质纱线纺制工程是一个棘手问题,混纺时需要考虑多方面问题。例如,买家关注价格和产量,染色商关注颜色和可染性,纺纱生产者关注长度和可纺性,地毯制造商关注纱线特性、地毯外观和性能。因此,羊毛的选择、混合、纺纱和整理对实现所有这些要求起着至关重要的作用。简言之,纱线工程可分为选毛与混合、纺纱与加捻、纱线整理三部分。

4. 新趋势

长期以来,高产和高速是地毯织造行业新发展的最重要因素。地毯织工想编织不同类型,并从一种品质快速转变为另一种品质。原材料的发展对这新趋势做出了积极贡献。新的染色技术、化合物和处理方式改进了纱线的质量,但也改善了原材料的选择。地毯织机的新发展和技术改进也带来了其他贡献。面对编织技术,高筘密、多色的地毯成为新趋势。现在,700 筘 / 米、8 色框的地毯已可编织。编织类似手工打结、无经纱的地毯背面,努力使地毯背面更清晰。生产具有像景、高低效果、高绒或蓬松效果、组合效果(割绒、圈绒、平纹效果组合)的地毯呈增长趋势。设计师希望创造力不受任何限制地绘制地毯。Axminster 织机是第一台 16 色框的高速全新织机。

5. 室内使用

住宅用地毯在世界各地制造,特别是美国和西欧。地毯制造中心正在印度次大陆和中国兴起。各中心的产品展现出了不同发展方式和风格特征。一般而言,美国地毯一般带有聚酰胺起绒图案和纹理绒圈、较长涤纶割绒的“萨克森”、更经济的聚丙烯绒结构。

6. 商业使用

商业用地毯通常用于人流密集场合。办公室地毯,尤其是现代开放式会议室、大型零售场所,采用紧密耐磨小绒圈结构的聚酰胺绒。地毯片使用特别流行。地毯片通常为 50 平方厘米或 45.7 平方厘米(18 平方英寸)的方格,特别适合大型多层设施。最初地毯片为简单被衬黏合毡的方块。款式由针刺地毯、平针圈簇绒、平针割簇绒甚至图案结构发展而来,每个阶段款式均展现了风格、奢华和精致。

Lesson 47   Interior textiles

Textiles do much to create the mood of an interior through their contributions to colour, texture, pattern and performance. They soften the contours of the furniture, finish walls and windows, and otherwise enhance the interior. They can be used to absorb sound and make a space more pleasant for work, relaxation, and human interaction. The construction technique used to produce a specific fabric has an impact on its cost, texture and appearance, and its suitability and performance for a specific end use.

1.Natural fibres for interior textiles

There are many interior applications for natural fibres as yarns, fabrics and non-wovens from tapestries, wall coverings, tablecloths, carpets, upholstery fabrics, bedsheets, window blinds and curtains to towels and mattresses. Natural fibrous materials are environmentally more user-friendly than man-made fibres. They guarantee optimal comfort of use and generate a positive influence on human physiology.

2.Synthetic fibres for interior textiles

All kinds of textile fibres (natural, synthetic, man-made and inorganic) are used in interior textiles. Their fields of application are quite varied, but usually synthetic fibres are used for carpets, curtains, coverlets, pillows and various decorative interior textiles. Many types of materials are used for interior textiles—yarns, woven, knitted, nonwoven fabrics, etc. The six most popular kinds of synthetic fibres used for interior textile manufacturing are polyamide, polyester, polypropylene, polyethylene, acrylic and fibres of the meta-aramid group. Manufacturing of interior textiles also requires non-flammable synthetic fibres, such as meta-aramids and polyamides. The chemical structure and properties of these fibres do not differ from analogous fibres used for clothing or other technical applications.

3.Surface design of fabrics for interior textiles

Surface design of textiles constitutes the appearance of the fabric surface in terms of its colour, texture, pattern. The surface of the fabric is an integrated part of the whole, especially in the case of interior textiles where the performance, durability, and other functional requirements of the fabric is of extreme importance. Consequently, the surface design should be seen, not just as an aesthetic packaging of the product, but also as a summary of the textile's quality; a convergence of the effect of all the finishes and treatments goes into making the textile with its visual qualities.

生词与词组

1.interior application 室内应用
2.tapestry [ˈtæpəstri]n. 挂毯
3.wall covering 墙布
4.tablecloth [ˈteɪblklɒθ]n. 桌布
5.window blind 百叶窗
6.towel and mattress 毛巾和床垫室内用品
7.coverlet [ˈkʌvələt]n. 床罩
8.pillow [ˈpɪləʊ] n. 枕头

译文

<h2 style="text-align:center">第 47 课　装饰用纺织品</h2>

纺织品通过其颜色、质地、图案和性能,对营造室内氛围起很大作用。它们软化了家具的轮廓,装饰了墙壁和窗户,也增强了室内效果。它们可吸音,也可为工作、休息、人际交流营造舒适环境。用于生产特定织物的构造技术会影响其成本、质地和外观,以及特定用途的适用性和性能。

1. 装饰用天然纤维

从挂毯、墙布、桌布、地毯、室内装饰织物、床单、百叶窗和窗帘到毛巾和床垫室内用品,多数为天然纤维的纱线、织物、无纺布。天然纤维材料比人造纤维更环保。它们保证了最佳的使用舒适度,对人体生理产生积极影响。

2. 合成纤维的室内纺织品

各种纺织纤维(天然、合成、人造和无机)用于室内纺织品。它们的应用领域非常广泛,但合成纤维通常用于地毯、窗帘、床罩、枕头等室内装饰纺织品。许多类型的材料用于室内纺织品,如纱线、机织、针织、无纺布等。聚酰胺、聚酯、聚丙烯、聚乙烯、腈纶和间位芳纶纤维是最流行的六种用于室内纺织品的合成纤维。室内纺织品也需使用不易燃的合成纤维,如间位芳纶和聚酰胺。这些纤维的化学结构和特性与服用或其他技术应用的同类纤维无区别。

3. 室内纺织品面料设计

纺织品的表观设计由织物表面的颜色、质地和图案组成。室内纺织品的织物表面为其整体一部分,织物的性能、耐用性和其他功能也极为重要。因此,表面设计应不仅被看作产品的美观包装,也是纺织品质量体现;所有整理和处理的效果都融合进纺织品感官质量。

Lesson 48　Medical textiles

1.The role of textile structures and biomaterials in healthcare

New generation of medical textiles is an important and growing field. The importance of medical textiles is determined by their excellent physical, geometrical, and mechanical qualities, such as strength, extensibility, flexibility, air, vapour and liquid permeability, availability in two- or three-dimensional structures, variety in fibre length, fineness, cross-sectional shape, etc. Nowadays, textile products are able to combine traditional textile characteristics with modern multifunctionality and this role is constantly evolving. Medical textiles should provide many specific functions depending on the scenario (healthcare monitoring or healing), application peculiarity, individuality of the patient and so on. Specialised materials with determined functions can be included in medical textiles, extending into multifunctional systems made from natural or/and manufactured (man-made) materials. The role of medical textiles and biomaterials is determined by their leading features, depending on the final application. Such materials could be bacteriostatic, anti-viral, non-toxic, highly absorbent, non-allergic, breathable, haemostatic, biocompatible and incorporating medications, and can also be designed to provide reasonable mechanical properties and comfort. A wide variety of textile structures can be used for medicine and healthcare: fibre (or filament), sliver, yarn, woven, nonwoven, knitted, crochet, braided, embroidered, composite materials, etc. Medical textiles also use materials like hydrogels, matrix (tissue engineering), films, hydrocolloids, and foams. The advantage is that the materials can be used as gels, films, sponges, foams, fibres, support matrices and in blends or combinations as well. Specialised additives with special functions can be introduced in advanced products with the aim of absorbing odours, providing strong antibacterial properties, reducing pain and relieving irritation. Nanofibres are used due to their unique properties such as high surface area to volume ratio, film thinness, nanoscale fibre diameter, porosity, and light weight.

2.Types of textiles and biomaterials for medical applications

Textiles for healthcare include fibres, filaments, yarns, woven, knitted, nonwoven materials, and articles made from natural and manufactured materials as well as products utilising such raw materials. In textile products, the fibres are the main conventional structural elements. The most important natural fibres for healthcare are cotton, silk and flax. These fibres are the oldest textile structures used in medical products. The manufactured fibres, which are applied in the healthcare sector, may be subdivided into organic and inorganic fibres. Organic fibres can be divided into two large groups based on natural and synthetic polymers. The entire spectrum of manufactured fibres from such polymers as polyester (PES), polyamide (PA), polypropylene (PP), viscose (CV) and polytetrafluoroethylene (PTFE) has found increasing application during the last decades. Polyethylene terephthalate (PET) as the most common fibre forming PES is also used. Besides these fibres, a variety of fibrous medical materials have been derived from natural polymers, such as alginate (ALG), polylactic acid (PLA), distinct types of collagen, etc.

3.Properties of medical textile products

The suitability of polymeric materials and textile products applied in the healthcare field is credited to their unique origin and properties enabling them for such diverse applications as hygiene, protection, therapeutic, non-implantable or implantable materials, extracorporeal devices, etc. Cellulose is the most widespread natural polymer, and is now the basic polymer used in the mass production of many materials for healthcare. Keratin and fibroin are structural biopolymers used for biomedical applications because of their useful properties including biodegradability and good biocompatibility. These biopolymers can be extracted from hair, wool, silk, nails, and feathers. Silk-fibroin has for centuries been used as a suitable raw material for manufacturing surgical threads. Materials obtained from natural silk-fibroin are characterised by good water vapour and air permeability, as well as high biocompatibility.

生词与词组

1.healthcare monitoring 保健监测

2.bacteriostatic [bækˌtɪrɪəˈstætɪk]*adj.* 抑菌的

3.anti-viral [ˈæntɪ ˈvaɪrəl]*adj.* 抗病毒的

4.non-toxic [nɒn ˈtɒksɪk]*adj.* 无毒的

5.non-allergic [ˌnɒnəˈlɜːdʒɪk]*adj.* 不过敏的

6.incorporating medication 含药物的

7.crochet [ˈkrəʊʃeɪ] *n.* 钩针编织品

8.embroidered [ɪmˈbrɔɪdəd]*adj.* 绣花的

9.hydrogel [ˈhaɪdrəˌdʒel]*n.* 凝胶

10.tissue engineering 组织工程

11.hydrocolloid [ˌhaɪdrəʊˈkɒlɒɪd]*n.* 水状胶质；水解胶体

12.sponge [ˈspʌndʒ]*n.* 人造海绵；海绵状物

13.specialised additive 专用添加剂

14.pain and relieving irritation 疼痛和缓解刺激

15.high surface area to volume ratio 高比表面积

16.nanoscale fibre diameter 纳米级纤维直径

17.alginate [ˈældʒɪneɪt]*n.* 海藻酸盐

18.therapeutic [ˌθerəˈpju:tɪk]*adj.* 治疗的

19.non-implantable or implantable 不可植入的或可植入的

20.extracorporeal device 体外装置

21.keratin and fibroin 角蛋白和丝素蛋白

22.nail [neɪl]*n.* 指甲

23.surgical thread 手术线

译文

第48课　医用纺织品

1. 纺织结构生物材料的医疗保健作用

新一代医用纺织品是重要且不断发展的领域。医用纺织品的重要性能取决于其优异的物理、几何和机械特性(如强度、延展性、柔韧性、气液渗透性、可获得二维或三维结构性、纤维长度范围、细度、截面形状等)。现在,传统的纺织品可与新功能性相结合,并且在不断进步。医用纺织品应根据设想(保健监测或愈合)、应用特性、患者的个体情况等进行功能设定。在医用纺织品中所包括的特定功能专用材料,已扩展到由天然或/和人造材料制成的多功能系统。医用纺织品和生物材料的作用取决于它们的主要特性及其最终用途。这些材料是抑菌、抗病毒、无毒、高吸水性、不过敏、透气、止血、生物相容和含药物的,还可设计成提供机械性能和舒适度的。

多种纺织结构可用于医药和保健:纤维的(或长丝的)、条的、纱线的、机织的、无纺布的、针织的、钩针的、编织的、绣花的、复合材料的等。医用纺织品还可用水凝胶、基质(组织工程)、薄膜、水胶体和泡沫等材料。这些材料的优势是可用作凝胶、薄膜、海绵、泡沫、纤维、支撑基质及其混合或组合。具有特殊功能的专用添加剂添加入先进产品,以达到吸收异味、强抗菌性能、减轻疼痛和减轻刺激等目的。纳米纤维因其独特的特性(如高比表面积、膜薄、纳米级纤维直径、孔隙率和质轻)而被使用。

2. 医疗保健用纺织结构生物材料的类型

医疗保健的纺织品包括纤维、长丝、纱线、机织、针织、非织造材料、天然和人造材料制成的产品以及这些产品的原材料。在纺织产品中,纤维是主要的结构单元。医疗保健用的天然纤维主要为棉、丝和亚麻。这些纤维是最早用于医疗产品的纺织材料。用于医疗保健的人造纤维可分为有机纤维和无机纤维。有机纤维分为天然和合成聚合物两大类。过去的几十年,聚酯(PES)、聚酰胺(PA)、聚丙烯(PP)、黏胶(CV)和聚四氟乙烯(PTFE)等全系列的聚合物纤维得到了快速应用。由 PES 生成的聚对苯二甲酸乙二醇酯(PET)作为最常见的纤维被使用。除了这些纤维之外,各种源自天然聚合物的医用纤维材料被应用,如海藻酸盐(ALG)、聚乳酸(PLA)、不同类型的胶原蛋白等。

3. 医用纺织产品的性能

聚合物材料和纺织产品适用于医疗保健领域,这归功于其独特的来源和特性,使它们可用于卫生、防护、治疗、非植入或可植入材料、体外装置等多种用途。纤维素是最普遍的天然聚合物,是目前大量生产的医疗保健材料中最基本的聚合物。因其具有独特的生物降解性和良好的生物相容性,角蛋白和丝素蛋白作为生物结构聚合物用于生物医学应用。这些生物聚合物可从头发、羊毛、丝绸、指甲和羽毛中提取。几个世纪以来,丝素蛋白一直是手术线的最佳原材料。从天然丝素蛋白中提取的材料具有良好的水汽渗透性、高生物相容性。

Lesson 49 Sportswear

1.Overview

Textile materials are used in all sports as sportswear, and in many games as sports equipment and sports footwear. Examples of sportswear are: aerobic clothing, football clothing, cricket clothing, games shorts, gloves, jackets, pants, shirts, socks, sweatshirts, swimwear and tennis clothing. Examples of sports equipment are: sails, trampolines, camping gear, leisure bags, bikes and rackets. Examples of sports footwear are: athletic shoes, football boots, tennis shoes and walking boots. The consumption of textile fibres and fabrics in sportswear and sporting related goods has seen a significant increase in the last decade.

Textile materials in various shapes and forms are being used in a wide range of applications in sportswear and sports equipment, and the manufacturers of these products are often at the forefront of textile manufacturing technologies for enhancing the properties of performance fabrics and sportswear in order to fulfil various types of consumer and market demands. The performance requirements of many sporting goods often demand widely different properties from their constituent fibres and fabrics, such as barrier to rain, snow, cold, heat and strength, and at the same time these textiles must fulfil the consumer requirements of comfort, drape, fit and ease of movement. Among the contributing factors responsible for successful marketing of functional sportswear and sporting goods have been advances made in the fibre and polymer sciences, and production techniques for obtaining sophisticated fibre, yarns and fabrics. The finishing and coating/laminating industries have done pioneering work in the area of developing these technologies towards the needs of the sportswear and sporting goods sectors resulting in unique products.

2.Design considerations in sportswear/footwear

Design requirements of performance sportswear have produced designers with skills and knowledge in graphics, textiles and fashion to conceive aesthetically pleasing and ergonomically viable ranges which take advantage of the latest advances in functional and "smart" textiles. Leading fashion designers have been quick to realize that the performance has actually become the aesthetics in sportswear. It is the fabrics and technology that set the trend. Incorporation of microfibres, breathable barrier fabrics, innovative stretch materials, intelligent textiles, interactive materials such as phase-change materials and shape-memory polymers, and wearable technology as a part of the functional design system in sportswear, will become routine in the product development process.

3.Innovations in fibres and textile materials for sportswear

The evolution of fibre developments has gone through the phases of conventional fibres, highly functional fibres and high-performance fibres. Polyester is the most common fibre used for sportswear. Other fibres suitable for active wear are polyamide, polypropylene, acrylics and elastanes. Wool and cotton fibres are still finding applications in leisure wear. Synthetic fibres can either be modified during manufacture, e.g., by producing hollow fibres and fibres with irregular cross-section, or be optimally blended with natural fibres to improve their thermo-physiological and sensory properties.

Synthetic fibres with improved UV resistance and having antimicrobial properties are also commercially available for use in sportswear. Improved fibre spinning techniques in melt spinning, wet spinning, dry spinning as well as new techniques such as gel spinning, bi-component spinning and microfibre spinning, have all made it possible to produce fibres, yarns and fabrics with unique performance characteristics suitable for use in sportswear and sporting goods. New technologies for producing microfibres have also contributed towards production of high-tech sportswear. By using the conjugate spinning technique, many different types of sophisticated fibres with various functions have been commercially produced, which has resulted in fabrics having improved mechanical, physical, chemical and biological functions. The technique of producing sheath/core melt spun conjugate fibres has been commercially exploited for producing added-value fibres.

4.Sportswear and comfort

In endurance sports, the performance of a sportswear is synonymous with its comfort characteristics. In sportswear for outdoor use, the clothing should be capable of protecting the wearer from external elements such as wind, sun, rain and snow. At the same time, it should be capable of maintaining the heat balance between the excess heat produced by the wearer due to increased metabolic rate on the one hand, and the capacity of the clothing to dissipate body heat and perspiration on the other.

5.Sportswear and protection from injury

For impact protection to be provided by the clothing or sports equipment such as protective helmets, it is necessary to use textiles and textile-based materials which possess high strength and durability as well as a high level of energy absorption. These materials are attached to the clothing in appropriate places depending on the sporting activity and the information available from injury risk analyses of different sports and games. A variety of textiles and textile composite structures are commercially available with the required mechanical properties of strength, impact resistance, abrasion resistance and tear strength for rugged outdoor and performance sports and games.

6.The sportswear and sports footwear industry

We have been noticing a strong trend towards the combination of multi-functionality of materials and fashion in many items of sportswear and leisure clothing. The result has been the emergence of many different types of high-tech fabrics and garment designs with some remarkable performance properties.

生词与词组

1.aerobic clothing 有氧运动服

2.cricket clothing 板球服

3.glove [glʌv]n. 手套

4.pants [pænts]n. 裤子

5.sweatshirt [ˈswetʃɜːt]n. 运动衫

6.tennis clothing 网球服

7.trampoline [ˈtræmpəliːn]n. 蹦床

8.camping gear 露营装备

9.leisure bag 休闲包

10.racket [ˈrækɪt]n. 球拍

11.athletic shoe 运动鞋

12.tennis shoe 网球鞋

13.barrier [ˈbæriə(r)]n. 栅栏；屏障

14.breathable barrier fabric 透气隔热面料

15.innovative stretch material 超弹材料

16.intelligent textile 智能纺织品

17.interactive material 交互材料

18.phase-change material 相变材料

19.shape-memory polymer 形状记忆聚合物

20.thermo-physiological 热生理性

21.sensory property 感官特性

22.antimicrobial [ˌæntɪmaɪˈkrəʊbɪəl]*adj.* 抗微生物的

23.microfibre spinning 超细纤维纺丝

24.conjugate spinning 共轭纺丝

25.added-value fibre 附加值纤维

26.endurance sport 耐力运动

27.metabolic rate 代谢速率

28.perspiration [ˌpɜːspəˈreɪʃ(ə)n]*n.* 汗水

29.helmet [ˈhelmɪt]*n.* 头盔

30.injury risk 受伤风险

31.tear strength 撕裂强度

32.rugged outdoor 崎岖的户外

译文

第 49 课　运动服装

1. 概述

　　所有运动的运动服、许多比赛的运动器材和运动鞋都使用纺织材料。运动服的例子有：有氧运动服、足球服、板球服、比赛短裤、手套、夹克、裤子、衬衫、袜子、运动衫、泳装和网球服。运动器材的例子有：帆、蹦床、露营装备、休闲包、自行车和球拍。运动鞋的例子有：运动鞋、足球鞋、网球鞋和步行靴。过去十年，运动服及运动相关商品中纺织纤维、织物的消耗量显著增加。

　　各种形状和形式的纺织材料被广泛应用于运动服和运动装备中，这些产品的制造商往往紧跟前沿纺织制造技术，以提升面料和运动服的性能来满足不同类型的消费者和市场需求。许多体育用品的性能需求通常要求其所构成纤维和织物具有截然不同的特性，如防雨、防雪、防寒、防热和强力，同时这些纺织品也必须满足消费者对舒适性、悬垂性、合身和便于活动的要求。功能性运动服和运动用品的成功营销的关键在于纤维和聚合物科学以及尖端纤维、纱线和织物的生产技术所取得的进步。在开发这些技术方面,整理和涂层/层压行业做了开创性的工作,以满足运动服和体育用品行业的需求,从而产生独特的产品。

2. 运动服/鞋类的设计要点

高性能运动服的设计需要设计师具备图形、纺织品和时尚方面的知识和技能,以利用功能性和"智能"纺织品的最新进展来构思美观且符合人体工程学的可行系列。著名时尚设计师很快意识到,性能实际上已成为运动装的美学。引领潮流的是面料和技术。超细纤维、透气隔热面料、超弹材料、智能纺织品、相变材料、形状记忆聚合物以及可穿戴技术作为运动服功能设计的一部分,成为产品开发的常规过程。

3. 运动纺织品的纤维和材料创新

纤维发展经历了常规纤维、高功能纤维和高性能纤维的阶段。涤纶是用于运动服的常见纤维。其他适用于运动服的纤维是聚酰胺纤维、聚丙烯纤维、腈纶和弹性纤维。羊毛和棉纤维也用于休闲装。合成纤维既可在制造过程中进行改性(如中空纤维和不规则横截面的纤维),也可与天然纤维混合,以改善它们的热生理性和感官特性。

具有抗紫外线性和抗微生物性的合成纤维已可购买并用于运动服。熔融纺、湿纺、干纺等纤维纺丝技术的进步,以及凝胶纺丝、双组分纺丝和微纤维纺丝等新技术,使具有独特性能且用于运动服和体育用品的纤维、纱线和织物的生产成为可能。新技术生产的微纤维也为高科技运动服提供原料。通过使用共轭纺丝技术,许多具有各种功能的复杂纤维开始商业化生产,从而改进了织物的机械、物理、化学和生物性能。皮芯熔纺复合纤维的生产技术已开始商业应用于生产高附加值纤维。

4. 运动服和舒适度

在耐力运动中,运动服的性能等同于其舒适性。在户外使用的运动服中,服装应该可保护穿着者免受外界因素的影响,如风、日、雨和雪。同时,它应能保持穿着者因新陈代谢增加产生的多余热量和服装散掉身体热量以及汗水的能力之间的热量平衡。

5.运动服和防护

服装或运动装备（如防护头盔）提供的冲击保护,须使用高强度和高耐用性以及高能量吸收的纺织品或纺织品基材。根据体育活动以及从不同运动和比赛的伤害风险分析中获得的信息,将这些材料贴在衣服的适当位置。市售的各种纺织品和纺织复合结构具有高强度、抗冲击性、耐磨性和撕裂强度等机械性能,适用于崎岖的户外和高性能运动和比赛。

6.运动服和运动鞋业

我们已经注意到许多运动服和休闲服的强劲趋势,材料与时尚的多功能性相结合。结果出现了许多不同类型且带有卓越性能的高科技面料和服装设计。

Lesson 50 Hygiene textile products

1.Introduction

Hygiene products form an important group of medical textiles. With the growing global population, longer life span, and improved hygiene and healthcare standards, textile materials used in the healthcare/hygiene sector have gradually taken on more important roles. Recently, the use of these products has also penetrated into the household sector. Owing to recent advancements in the polymer, fiber and textile engineering fields, the use of textile materials in this sector has witnessed a tremendous growth.

2.Applications of hygiene products

Hygiene products include both disposable and non-disposable items that are mainly used in hospitals, such as antimicrobial textiles, towels, diapers, sanitary napkins, tampons, panty shields, wipes, incontinence products and so on. Based on their application, these can be broadly classified into the following categories:

• Wipes. Baby wipes are commonly based on a nonwoven substrate material that is coated or impregnated with a liquid lotion, packaged in such a way that the wipes are dispensed as required. The absorbency of the nonwoven substrate is important in achieving good non-linting and cleaning performance from the wipes.

• Baby diapers. Baby diapers provide an effective absorbent

structure to receive, absorb and retain urine and waste products from babies. The products should deliver absorption/retention functionality in such a way as to prevent irritation of the baby's skin and contamination of the baby's clothing, and be capable of disposal after use.

•Feminine hygiene products. The requirement of these absorbent products is to absorb and retain menstrual fluid discharges. The products are in intimate contact with the user, hence they must be without skin irritant tendencies, and provide containment and absorption without leakage.

•Adult diapers and incontinence pads. Adults who have to remain on duty for long durations (such as nonstop drivers and astronauts who cannot reach the toilet on time) and those who do not have control over their continence, find adult diapers very useful. They are worn like an under garment. These absorbent products can be durable (cloth type) or disposable. The disposable category is hygienic, comfortable and multi-layered. The interior layer is highly absorbent, has high wicking tendency and can retain fluid. The exterior layer is composed of a waterproof material. Hence, they absorb the fluid and at the same time keep the skin dry. Recent developments are hybrid reusable diapers and ultra thin diapers. Hybrid reusable diapers are made up of a material that is fully flushable and compostable to give extra care and comfort to adults as well as to babies. Ultra thin diaper manufacturing is possible due to the emergence of superabsorbent technology. These are highly functional and comfortable.

3.New technology to improve the performance of hygiene products

The films, tapes, adhesives, fluff pulps, superabsorbents and other materials that comprise baby diapers, feminine hygiene items and other disposable hygiene products are constantly being researched, and manufacturers continue to be challenged with making more sophisticated and advanced products to feed consumers' craving for better designs and performance. In recent years, improvements in diapers, in particular, have included more stretchable waistbands, improved leg cuffs, thinner and more absorbent cores, more textile-like backsheets, and environmentally responsive products. Many aspects must be considered in the design of the hygiene products (and raw materials or components of them) to improve their performance as well as aesthetics.

生词与词组

1.non-disposable [ˈnɒndɪsˈpəʊzəbl]*adj.* 非一次性的

2.antimicrobial [ˌæntɪmaɪˈkrəʊbiəl]*adj.* 抗菌的

3.diaper [ˈdaɪəpə(r)]*n.* 尿布

4.sanitary napkin 卫生巾

5.panty shield 护垫

6.incontinence product 失禁产品

7.liquid lotion 液体乳液

8.non-linting [ˈnɒnlɪntɪŋ]*adj.* 不掉毛的

9.baby diaper 婴儿纸尿裤

10.menstrual [ˈmenstruəl]*adj.* 月经的

11.skin irritant tendency 皮肤刺激倾向

12.astronaut [ˈæstrənɔːt]*n.* 宇航员

13.wicking tendency 高芯吸能力

14.flushable [ˈflʌʃəbl]*adj.* 可冲洗的

15.compostable [kɒmˈpɒstəbl]*adj.* 可堆肥的；可降解的

16.superabsorbent [ˌsuːpərəbˈsɔːbənt]*n.* 高吸水性树脂

17.fluff pulp 短纤浆

18.feminine hygiene item 女性卫生用品

19.disposable hygiene product 一次性卫生用品

20.stretchable waistband 可伸缩腰带

21.leg cuff 腿箍

22.textile-like backsheet 类似纺织品的底片

译文

第 50 课　卫生用纺织品

1. 介绍

卫生用品是医用纺织品的重要组成。随着全球人口的增长、寿命的延长以及卫生和保健标准的提升,用于保健/卫生领域的纺织材料正逐渐成为重要的角色。近期,这些产品也进入了家居领域。由于聚合物、纤维和纺织工程领域的发展,应用在该领域的纺织材料已大幅攀升。

2. 卫生用品的应用

卫生用品包括一次性和非一次性用品,其主要用于医院,如抗菌纺织品、毛巾、尿布、卫生巾、卫生棉条、护垫、湿巾、尿失禁产品等。按照它们用途,其大致分为以下几类:

·湿巾。婴儿湿巾通常是非织造基材,该材料涂覆或浸渍乳液,按湿巾使用需求进行包装。非织造材料的吸附性对湿巾的防粘和清洁性能至关重要。

·婴儿纸尿裤。婴儿纸尿裤提供了一种有效的吸收结构来接纳、吸收和保留来自婴儿的尿液和废物。该产品带有吸收/保留功能以防止刺激婴儿皮肤和污染衣服,并可用后丢弃。

·女性卫生用品。 这些吸收品可吸收和保留排出的月经经液。该品与使用者密切接触,因此必须不刺激皮肤,并可收纳和吸收而不渗漏。

·成人尿布和失禁垫。 纸尿裤适用于长时间值班的成年人(如无法准时上厕所的司机和宇航员)以及无法控制排泄的成年人,如内衣一样穿着。这些吸收品是耐用的(布型)或一次性的。一次性用品应卫生、舒适,为多层结构。该品内层为高吸收材质,具有高芯吸能力,可保留液体。外层由防水材料组成。 因此,它们可吸收液体,保持皮肤干燥。最近的研制成果是混合型可重复使用的尿布和超薄尿布。混合型可重复使用的尿布由完全可冲洗和可堆肥的材料制成,为成人和婴儿提供额外的护理。由于超吸收技术的出现,超薄尿布的制造成为可能。这些产品都功能强大且舒适。

3. 提高卫生用品性能的新技术

　　婴儿纸尿裤、女性卫生用品和其他一次性卫生用品中的薄膜、胶带、黏合剂、短纤浆、超吸收剂和其他材料不断被研究，制造商不断探索更复杂和先进的产品以满足消费者对更佳设计和更高性能的需求。近年来，特别是尿布的革新，包括有更高拉伸性的腰带、改进的腿箍、更薄和更具吸收性的芯层、更像纺织品的底片和更环保产品。卫生产品的设计中必须考虑多方面(原材料或组件)，以提高其性能和美感。

Lesson 51 Coated and laminated textiles

1.Definitions

There are two definitions of a "coated fabric". The first one is, "a material composed of two or more layers, at least one of which is a textile fabric and at least one of which is a substantially continuous polymeric layer. The layers are bonded closely together by means of an added adhesive or by the adhesive properties of one or more of the component layers". The second definition is, "a textile fabric on which there has been formed in situ, on one or both surfaces, a layer or layers of adherent coating material". This second definition is regarded as the definition which most closely describes coated fabrics. The first quoted definition could also be applied to a laminated fabric. A "laminated fabric", also sometimes called a "bonded fabric", is considered to be different from a coated fabric, in that the layers are already pre-prepared and the second material can be a film, another fabric, or some other material.

2.Properties

Coated fabrics are engineered composite materials, produced by a combination of a textile fabric and a polymer coating applied to the fabric surface. The polymer coating confers new properties on the fabric, such as impermeabilty to dust particles, liquids and gases, and it can also improve existing physical properties, such as fabric abrasion. The fabric component generally determines the tear and tensile strength, elongation and dimensional stability, while the polymer mainly controls the chemical properties, abrasion resistance and resistance to penetration by liquids and gases. Many properties, however, are determined by a combination of both these components, and both base-fabric and polymer must be carefully selected by a thorough consideration of the properties.

3.Fabric finishing

A process related to fabric coating is "fabric finishing", where a chemical or a polymer covers only the yarns making up the fabric without closing the gaps in between. In fabric coating, the small gaps in between the individual yarns are covered to varying degrees. In the case of a waterproof coating, the gaps are fully "bridged" by the coating, the polymer forming a continuous layer on the fabric surface. In other coated fabrics, the gaps may not be fully covered, and these fabrics will be porous to varying degrees. Fabric finishes are usually applied by a "dip and squeeze" process known as padding or impregnation, which is widely used in textile factories. The need for distinction between "coating" and "finishing" is again largely academic—the two merge in certain instances, because repeated impregnation with some polymers can produce a continuous coating on a fabric.

4.Fabric lamination

Polymer materials, which may not be easily formulated into a resin or a paste for coating, can be combined with a fabric by first preparing a film of the polymer, and then laminating it to the fabric in a separate process. There are various techniques and several different types of adhesive and machinery used in the lamination process. Producing an adhesive bond, which will ensure no delamination or failure in use, requires lamination skills and knowledge of which adhesive to use. It is generally relatively simple to produce a strong enough bond, while the challenge is to preserve the original properties of the fabric and to produce a flexible laminate with the required appearance, handle and durability.

5.The technical scope of coated and laminated textiles

The textile coating industry is heavily dependent on the polymer and speciality chemical industries for polymers, additives, process aids, and many other chemicals necessary to produce the specified customer properties at the quality standard and durability demanded. The manufacturing procedures used in textile coating and lamination are related to similar processes in the film and paper industries; both films and papers are coated with polymers and laminated to other materials, and in fact the dividing line between paper and some nonwoven fabrics is quite diffuse. The nonwoven fabric industry is possibly the biggest single user of many polymer resins as binders. Textile coating uses some polymers used in the paint industry, such as acrylics and polyurethanes, and in fact some analysts now refer to the paint industry as the coating industry. An analogy has been made to painting; the best results are obtained in textile coating by the application of a few thin layers, rather than one thick layer—just like painting.

生词与词组

1.waterproof coating 防水涂层
2.dip and squeeze 浸压
3.impregnation [ˌɪmpregˈneɪʃ(ə)n]n. 浸渍
4.coating [ˈkəʊtɪŋ]n. 涂层
5.resin [ˈrezɪn]n. 树脂
6.delamination [diːˌlæməˈneɪʃ(ə)n]n. 分离成层；脱层
7.polyurethane [ˌpɒliˈjʊərəθeɪn]n. 聚氨酯

译文

第 51 课　涂层和层压纺织品

1. 定义

"涂层织物"有两种定义。第一个是"由两层或多层的材料组成,至少一层是纺织织物,至少一层是连续的聚合物层。这些层通过添加黏合剂或由一个或多个组成层的黏合特性黏合在一起"。第二个定义是"一种纺织织物,在其一面或两面上涂覆一层或多层黏附材料"。第二个定义被认为是最贴切的涂层织物定义。第一个定义可指层压织物。"层压织物"有时也称为"黏合织物",被认为不同于涂层织物,因为这些层已经预先准备好,第二种材料可以是薄膜、另一种织物或某种其他材料。

2. 性能

涂层织物是工程复合材料,由纺织织物和涂在织物表面的聚合物层组成。聚合物涂层赋予织物新的特性(如灰尘颗粒、液体和气体的不渗透),也可改善现有的物理特性(如织物耐磨性)。织物构成通常决定了其撕裂和拉伸强度、伸长率和尺寸稳定性,而聚合物决定了其化学性能、耐磨性和水气渗透性。然而,许多特性是由两种成分共同决定的,必须全面考虑其特性来选择基础织物和聚合物。

3. 织物整理

与涂层织物相关的工艺是"织物整理",其是化学物质或聚合物覆盖织物内的纱,而不封闭它们之间的间隙。在涂层织物中,纱线间的间隙会被不同程度地覆盖。在防水涂层中,间隙被涂层完全覆盖,聚合物在织物表面形成连续层。在其他涂层织物中,间隙可能不完全覆盖,这些织物会带有不同程度的孔隙。织物整理通常由浸压工艺(称为"填充"或"浸渍")实现,这在纺织工厂广泛使用。"涂层"和"整理"的区分多为学术性的——一些情况下二者可合并,因为重复浸渍聚合物可在织物上形成连续涂层。

4.织物层压

聚合物材料不易制成树脂或涂层的糊料,可先制备成聚合物薄膜,再单独将其压合到织物上。层压工艺涉及各种技术、黏合剂和机械。制造不分层、不失效的黏合剂,需熟悉层压技能和黏合剂知识。制备有足够黏性的黏合剂相对简单,而挑战在于保持织物的原始特性的同时,生产具有所需外观、手感和耐用性的柔性层合材料。

5.涂层和层压纺织品的技术范围

纺织涂层行业严重依赖于聚合物和特种化学品行业的聚合物、添加剂、加工助剂以及其他按照质量标准和耐久性要求制造的特定性能的化学品。纺织涂层和层压的制造工艺与薄膜、造纸工业的工艺相类似;薄膜和纸都可涂覆聚合物,层压其他材料,事实上纸和一些无纺布的区分是模糊的。无纺布行业是使用多种聚合物树脂作为黏合剂的最大单一用户。纺织涂层使用一些用于涂料行业的聚合物,如丙烯酸树脂和聚氨酯,事实上一些分析家已将涂料行业称为"涂层行业"。与绘画相比,纺织品涂层是涂覆了几层薄料而不是一厚层来获得最佳效果,其就像绘画一样。

Lesson 52　Smart technology for textiles and clothing

　　Such textile materials and structures are becoming possible as the result of a successful marriage of traditional textiles/clothing technology with material science, structural mechanics, sensor and actuator technology, advanced processing technology, communication, artificial intelligence, biology, etc. We have always been inspired to mimick nature in order to create our clothing materials with higher levels of functions and smartness. The development of microfibres is a very good example, starting from studying and mimicking silk first, then creating finer and, in many ways, better fibres. However, up to now, most textiles and clothing have been lifeless. It would be wonderful to have clothing like our skin, which is a layer of smart material. Although the technology as a whole is relatively new, some areas have reached the stage where industrial application is both feasible and viable for textiles and clothing.

　　Many exciting applications have been demonstrated worldwide. Extended from the space programme, heat generating/storing fibres/fabrics have now been used in skiwear, shoes, sports helmets and insulation devices. Textile fabrics and composites integrated with optical fibre sensors have been used to monitor the health of major bridges and buildings.

　　The first generation of wearable motherboards has been developed, which has sensors integrated inside garments and is capable of detecting injury and health information of the wearer and transmitting such information remotely to a hospital. A friendly and comfortable wearable computer or a wearable mobile phone is integrated into some form of apparel. The Philips ICD+ can be viewed as the first generation of smart clothing because they integrate mobile phones and music players, which try to enhance the "organizer" functions of clothes. The keypad is placed on the sleeve of the jacket and a microphone is discretely placed on the collar. Shape memory polymers have been applied to textiles in fibre, film and foam forms, resulting in a range of high performance fabrics and garments, especially sea-going garments. Fibre sensors, which are capable of measuring temperature, strain/stress, gas, biological species and smell, are typical smart fibres that can be directly applied to textiles.

Conductive polymer-based actuators have achieved very high levels of energy density. Clothing with its own senses and brain, like shoes and snow coats which are integrated with Global Positioning System (GPS) and mobile phone technology, can tell the position of the wearer and give him/her directions. The soldiers may be connected to navigation systems via wearable computers to guide them through difficult terrain and unknown areas. If necessary, medical advice can be given or treatment administrated anywhere, not just in a hospital, thereby leading to more mobility as well as to more efficient and effective health services. Biological tissues and organs, like ears and noses, can be grown from textile scaffolds made from biodegradable fibres. Integrated with nanomaterials, textiles can be imparted with very high energy absorption capacity and other functions like stain proofing, abrasion resistance, light emission, etc.

生词与词组

1.actuator [ˈæktʃʊˌeɪtə]*n.* 执行器
2.artificial intelligence 人工智能
3.mimick [ˈmɪmɪk]*v.* 模仿
4.skiwear [ˈskiːˌweə]*n.* 滑雪服装
5.sports helmet 运动头盔

6.insulation device 绝缘装置

7.motherboard [ˈmʌðəbɔːd]*n.* 主板

8.sensor [ˈsensə]*n.* 传感器

9.microphone [ˈmaɪkrəfəʊn]*n.* 麦克风

10.sea-going garment 航海服装

11.conductive polymer-based actuator 基于导电聚合物的执行器

12.navigation [ˌnævɪˈgeɪʃ(ə)n]*n.* 导航

13.terrain [təˈreɪn]*n.* 地带；地形

14.nanomaterial [ˌnɒnməˈtɪərɪəl]*n.* 纳米材料

15.light emission 发光

译文

第 52 课　纺织品和服装的智能化

传统纺织／服装技术与材料科学、结构力学、传感器和执行器技术、先进加工技术、通信、人工智能、生物学等成功结合的纺织材料和结构正在实现。我们已经从模仿自然中得到启发，以创造具有更高功能和智能水平的服装材料。微纤维的发展就是一个很好的例子，首先是研究和模仿丝绸，再创造出更细、更独特的纤维。然而，截至目前，多数纺织品和服装缺乏活力。如果衣服能像我们的皮肤一样，是一层智能材料就太棒了。整体上该技术在纺织服装行业相对较新，一些领域已可工业化应用。许多振奋人心的应用已在世界范围内展示。太空计划衍生而来的发热／储热纤维或织物已用于滑雪服、鞋子、运动头盔和绝缘设备。集成了光纤传感器的纺织物和复合材料已被用于监测桥梁、建筑物的状况。

第一代可穿戴主板已诞生，传感器被集成在服装内，可监测装带者的伤害和健康信息，并将这些信息远程传输到医院。一款健康、舒适的可穿戴电脑或手机集成在某种服装中。飞利浦 ICD+ 被认为是第一代智能服装，它集成了手机和音乐播放器，试图增强衣服的"控制"功能。键盘设置于夹克的袖子上，麦克风分散在衣领上。形状记忆聚合物以纤维、薄膜和泡沫形式应用于纺织品，从而产生了一系列高性能织物和服装，尤其是航海服装。纤维传感器是典型的智能纤维，其可测量温度、张力／压力、气体、生物种类和气味，可直接应用于纺织品。

聚合物基质的导电执行器已可实现非常高的能量密度。拥有自我的感觉和大脑的服装，如整合了全球定位系统（GPS）和手机技术的鞋子和防雪外套，其可说出穿着者的位置并为他／她指明方向。士兵可通过可穿戴计算机连接到导航系统，引导他们穿越困难的地形和未知的区域。如需要，可在任何地方提供医疗建议或进行治疗，而不仅仅在医院，从而带来更多的可移动的以及更快捷、有效的卫生服务。生物组织和器官（如耳朵和鼻子）可从由生物降解的纤维制成的纺织支架上生长。

Lesson 53   Comfort of the garment

1.Introduction

As a key parameter of clothing, comfort of the garment is the complex effect of textile properties which are basically dependent on the chemical structure and morphology of the constituent fibers. Comfort properties of textile products such as yarns, fabrics, mats and any other product that is used for wearing purposes embrace different mechanical properties as well as heat and moisture transfer (or isolation).

Comfort can be considered as the consistency of the clothing with the human surroundings or environment determined by human pleasantness or relief. This consistency includes physical aspects such as heaviness, thickness, thermal transmission or heat transfer, air permeability, moisture absorbency and moisture diffusion, handle, drape as well as aesthetic aspects such as colour, luster, fashion, style, and fit and in addition to these aspects we have to consider the subjectivity of the user, culture, etc.

As comfort is definitely an individualistic sense, it is very difficult to define, design or determine it. Discomfort sense, which is the opposite point of comfort, may be easier to define using terms of prickle, hot (cold), tight, moist, etc., that happen when we are consciously aware of the unpleasantness of the worn clothes. Two major types of discomfort sensation are psychological discomfort and physiological discomfort.

Researcher has mentioned thermo-physiological wear comfort and skin sensational wear comfort as two aspects of wear comfort of clothing. The former concerns the heat and moisture transport properties of clothing and the way that clothing helps to maintain the heat balance of the body during various levels of activity, and the latter concerns the mechanical contact of the fabric with the skin, its softness and pliability in movement and its lack of pricking, irritation and cling when damp. However, the major definition of clothing comfort describes it as a decision made by the brain according to psychological, physical and thermo-physiological condition of the body. Psychological comfort is mainly related to the subjectivity of the wearer and aesthetic parameters such as the latest fashion trend, colour harmony, fit, and acceptability in society.

Comfort properties of fibers

For thousands of years, the presence of textile products in the life of men was limited by the inherent qualities available naturally: cotton, wool, silk, linen, hemp, ramie, jute and many other natural resources. Since 1910, with the production of rayon as "artificial silk", man-made fibers started to be developed and to be used. Nowadays man-made fibers are found in different features of life with countless applications, from apparel, sports clothes and furnishings to industrial, medical, aeronautics and energy. The main advantage of the commercial man-made fibers is their low cost with respect to natural fibers. Considering their remarkable mechanical and chemical properties, modern life cannot survive without man-made fibers. However in the field of comfort, man-made fibers can not overcome the natural fibers. The behavior of fabric is affected by chemical and physical properties of its constituent fibers, fiber content, physical and mechanical characteristics of its constituent yarns, and the finishing treatments which are applied on it. Cost, quality, care and comfort are the essential properties that the customer considers before buying the cloth, all of which are different aspects of comfort.

The type of fiber is the most crucial specification which determines important properties such as strength, durability, handle, elasticity, dyeability, luster, friction properties, moisture absorbance, heat isolation and abrasion resistance. Fiber type is the most effective parameter in defining the comfort of the end product. Fiber content is the ratio of presentation of different types of fibers in textile production. This factor determines the moisture absorbency which in turn affects the thermal balance of the product, the moisture vapor and liquid permeability, the durability and the electrostatic properties in particular. Natural fibers are generally believed to provide better comfort sense. This is an old idea that comes from the higher ability of natural fibers in providing better moisture absorbency, heat isolation, handle, luster, etc. Synthetic fibers are mostly deficient in providing warmth, adequate bulkiness, soft feeling, thermal isolation and moisture absorbency due to their physical properties and chemical structure. For example in terms of thermal comfort, an ideal cloth should have high thermal resistance for cold protection, low water vapor resistance to be efficient in heat transfer under a mild thermal stress condition, and has to have rapid liquid transport to eliminate unpleasant tactile sensations due to water under a high thermal stress condition. Compared to cotton products, polyester fabric has lower water vapor resistance, but is inferior in both thermal resistance and liquid water transport. However, synthetic fibers have high strength, durability, dimension stability, abrasion resistance coupled with their thermoplastic properties and good resistance against heat. All these properties motivate textile producers to employ them in apparel products, especially considering their low cost. Nowadays, the growth of knowledge of fiber science and technology has come in the service of fiber designing. Using the technology of fiber production it is possible to produce the synthetic fibers with a cross-section of natural fibers. It is also possible to produce modified synthetic fibers which present closer behavior to natural fibers and it is possible to employ special finishing treatments to modify the properties of produced synthetic fibers according to the preferable behavior of natural fibers.

3.Comfort properties of yarns

Yarn properties are first of all created by physical and chemical properties of their constituent fibers: the chemical nature of the fibers, surface tension, fiber diameter and cross-section. However, the spinning technique, yarn linear density, pore size in the yarn, the distribution of pore size and blend ratio are the other parameters influencing the properties of yarns such as strength, bending rigidity, evenness, frictional properties, wickability, thermal insulation, liquid vapor permeability, air permeability and the properties of fabrics and clothes which are made of them. Many researchers have focused on the effects of yarn on the comfort properties of clothes, but mainly on the effects of yarn spinning techniques, fiber content ratio, and yarn structure.

4.Comfort properties of fabric structures

Cloth is made from fabrics which are even knitted (interlocked loops), woven (interlacing threads), or nonwoven (matted fibers). Fabric physical properties create the comfort characteristics of the cloth. The major properties are fabric density, porosity, bulkiness, thickness, structure and pattern.

生词与词组

1.human pleasantness or relief 人类的愉悦或轻松感
2.moisture diffusion 水分扩散
3.prickle ['prɪkl]v. 刺痛
4.discomfort sensation 不适感
5.psychological discomfort 心理不适

6.physiological discomfort 生理不适

7.thermo-physiological wear comfort 热生理穿着舒适度

8.skin sensational wear comfort 皮肤感触穿着舒适度

9.aeronautics [ˌeərəˈnɔ:tɪks]n. 航空学

10.tactile sensation 手感

11.wickability [ˌwɪkəˈbɪləti]n. 芯吸性

12.interlacing thread 交织线

13.matted fiber 纠缠的纤维

译文

第53课　服装舒适性

1. 介绍

　　作为衣服的关键指标,服装的舒适是复杂的纺织品性能,其取决于纤维的化学结构和形态。纺织品(如纱线、织物、垫子以及其他穿着用的产品)的舒适性包括不同的机械特性以及热湿气传递(或隔绝)。

　　舒适被认为与服装和人体间的环境相一致,其由人类的愉悦或轻松感决定。这种一致性包括物理方面(如重量、厚度、热传递或热转移、透气性、吸湿性和放湿性、手感、悬垂性)、美学方面(如颜色、光泽、时尚、款式、合身)。除此之外,我们还必须考虑用户的主观性、文化等。

　　由于舒适是一种个人的绝对感觉,很难定义、设计或确定它。不适感是舒适的反意,用刺痛、热(冷)、紧、潮湿等术语更易于定义,当我们有意识地意识到所穿衣服的不愉悦时,就会出现这种情形。

　　研究者已将服装穿着舒适度分为热生理穿着舒适度和皮肤感触穿着舒适度两个方面。前者与服装的热和湿传导特性以及在各种活动期间服装维持身体热平衡的方式相关,后者与织物和皮肤间的机械接触、它活动的柔软度和柔韧性相关,在潮湿时不会刺痛、刺激和紧贴。然而,服装舒适度的概念是由大脑根据人体的心理、生理和热生理条件做出的决策。心理舒适度主要与穿着者的主观性和审美指标(如最新流行趋势、色彩和谐、合身、社会接受度等)有关。

2. 纤维的舒适性

几千年来,人们生活中的纺织品受到天然获取材料(如棉花、羊毛、丝绸、亚麻、大麻、苎麻、黄麻和许多其他自然资源)的固有品质的限制。1910 年,随着人造棉作为"人造丝"的生产,人造纤维开始被开发和使用。现在,人造纤维在生活的各个方面都有应用,从服装、运动服和家具到工业、医疗、航空和能源。商用人造纤维的主要优点在于其与天然纤维相比成本低。基于其优良的机械和化学特性,现代生活离不开人造纤维。在舒适性方面,人造纤维无法超越天然纤维。织物的行为受其内部纤维的化学和物理特性、纤维含量、纱线的物理和机械特性以及对其所施加的整理的影响。价格、质量、护理和舒适度是客户在购买布料之前考虑的基本属性,所有这些都是舒适度的不同方面。

纤维的种类是最关键的规格,它决定了强度、耐久性、手感、弹性、染色性、光泽、摩擦性、吸湿性、隔热性和耐磨性等重要性能。纤维类型是定义终端产品舒适度的最有效参数。纤维含量是纺织品中不同类型纤维的占比。该因素决定了吸湿性,特别影响了产品的热平衡、水气和液体渗透性、耐久性和静电特性。通常认为天然纤维更为舒适。这是一个老观念,其源于天然纤维可提供更好的吸湿性、隔热性、手感、光泽等性能。合成纤维由于其物理特性和化学结构,大多在保暖性、蓬松度、柔软感、隔热性和吸湿性方面存在不足。例如,在热舒适性方面,理想的布料应具有高热阻隔性以进行防寒,低水汽阻力以在温和的热应力条件下有效导热,并具有快速的导湿能力以消除高热应力条件下由水带来的不悦触感。与棉制品相比,涤纶织物的水汽阻力较低,但热阻和排湿均较差。然而,合成纤维具有高强、耐用性、尺寸稳定、耐磨以及热塑性和良好的耐热性。所有这些特性都促使纺织生产商将它们用于服装产品,特别是它们较低的价格。现在,先进的纤维科学和技术已经服务于纤维设计。使用纤维生产技术可生产天然纤维横截面的合成纤维,也可生产出性能更接近天然纤维的改性合成纤维,还可采用特殊的整理方法制造具有天然纤维卓越性能的合成纤维。

3. 纱线的舒适性

纱线的特性源于其组成纤维的物理和化学特性:纤维的化学性质、表面张力、纤维直径和横截面。然而,纺纱技术、纱线线密度、纱线孔隙尺寸、孔隙尺寸分布和混纺比是影响纱线性能(如强度、弯曲刚度、均匀度、摩擦性能、芯吸性、隔热性、水汽渗透性、透气性以及其织物和衣服的性能)的其他参数。许多研究人员关注纱线对衣服舒适性的影响,但主要是纱线纺纱技术、纤维含量比和纱线结构的影响。

4.织物结构的舒适性

衣服由针织(互锁线圈)、机织(交织线)或非织造(纠缠的纤维)布料制成。织物的物理特性缔造了衣服的舒适特性,主要特性是织物密度、孔隙率、蓬松度、厚度、结构和图案。

Lesson 54   Subjective assessment of clothing appearance

1.Introduction

Clothing appearance is one of the most important aspects of clothing quality. People do have a reasonably common notion or concept of what is good or bad appearance. With the exception of some deliberate use of "puckered" or "wrinkled" surfaces, a nicely smooth and curved garment surface is regarded as desirable. Clothing is often discarded because of an unacceptable deterioration or change in appearance, including loss of shape, surface degradation, colour change, change in handle and pilling. The evaluation of clothing appearance is critical to product development and quality assurance. Subjective visual assessment is still the industrial norm because of the limitations of the many objective measurement systems. Visual assessments can be carried out on the materials and components of clothing as well as on the overall appearance of the clothing.

2.Assessment of fabric surface smoothness

•Assessment of fabric wrinkle recovery

A large number of techniques and methods exist for assessing fabric wrinkle appearance and recovery. One of the factors which influences clothing appearance is the ability of fabrics to recover from induced wrinkles or to retain a smooth surface appearance after wear and repeated laundering. The method often used in industry to evaluate the wrinkle recovery of a fabric is AATCC Test method 128 "Wrinkle Recovery of Fabrics".

•Assessment of pilling

The appearance quality of clothing are also influenced by the fabric propensity to surface fuzzing and pilling. Pills are developed on a fabric surface in four main stages: fuzz formation, entanglement, growth and wear-off. The formation of pills and other related surface changes (e.g., fuzzing) on textile fabrics during garment wear can create an unsightly appearance. This is a particularly serious problem with some synthetic fibres, where the strong synthetic fibres anchor the pills to the fabric surface, not allowing them to fall off as is the case with the weaker natural fibres. The pilling resistance of fabrics is normally tested by simulated wear through tumbling, brushing or rubbing on a laboratory testing machine. The specimens are then visually assessed by comparison with visual standards (either actual fabrics or photographs) to determine the degree of pilling on a scale ranging from 5 (no pilling) to 1 (very severe pilling).

•Surface smoothness after repeated laundering

AATCC Test Method 12419 is designed for evaluating the appearance, in terms of smoothness, of flat fabric specimens after repeated home laundering. The test procedure and evaluation method are almost the same as in the methods mentioned above, except for the difference in specimen preparation and standard replicas.

•Assessment of seam appearance

Visual assessment of seam appearance is conducted by comparing the seams with photographic standards under standard viewing conditions. The American Association of Textile Chemists and Colourists (AATCC), American Society for Testing Materials (ASTM), International Organisation for Standardisation (ISO) and Japan Industrial Standard (JIS) have established respective standards and procedures for visual assessment.

•Assessment of crease retention

To maintain good garment appearance, the pressed-in creases in garments (especially in trousers) should be retained after repeated home laundering. AATCC Test Method 88C27 is designed for evaluating the quality of crease retention in the fabric. The principle of the method is to subject creased fabric specimens to standard home laundering practices and then rate the appearance of specimens in comparison with appropriate reference standards under a standard lighting and viewing area. A choice is provided of hand or machine washing, alternative machine wash cycles and temperatures, and alternative drying procedures. Three representative fabric specimens cut parallel to the fabric length and width, are prepared, pressed and rated, respectively. The AATCC crease retention replicas are in five grades.

•Assessment of appearance retention of finished garments

Garment appearance may deteriorate due to poor fabric dimensional stability and pressing performance, poor workmanship during garment manufacture and unfavourable conditions during transport. This problem is especially acute for wool garments. Consequently, the International Wool Secretariat, Japanese branch proposed a test method for assessing the appearance retention of men's suits after final pressing and prior to sale. The principle of the test is to expose garments to certain temperature and humidity conditions for a period of time and then to check the changes in appearance afterwards.

生词与词组

1. visual assessment 视觉评估
2. fuzzing ['fʌzɪŋ]n. 起毛
3. fuzz formation 绒毛形成
4. wear-off 脱落
5. anchor ['æŋkə(r)]v. 停泊；使固定
6. tumbling ['tʌmblɪŋ]n. 翻滚
7. brushing or rubbing 刷磨或揉搓

8.laboratory testing machine 实验室试验机

9.trousers [ˈtraʊzəz]*n.* 裤子

10.crease retention 折痕保留

译文

第54课　服装表观评估

1.介绍

　　服装外观是服装质量最重要的方面之一。人们对外表的好坏有一个相当普遍的观念。除了一些特意"褶皱"或"起皱"的表面外,顺滑有弧度的服装表面被认可。服装经常因外观变差而被丢弃,其包括形状消失、表面风化、变色、手感变化和起球。服装外观的评估对产品开发和质量认证至关重要。由于客观测量系统的局限性,主观视觉评估仍是行业标准。视觉评估可对服装的材料和构成以及服装的整体外观进行评价。

　　2.织物表面光滑度评估

　　·织物折皱回复评估

　　有多种评估织物折皱回复的技术方法。影响服装外观的因素之一是织物在磨损和反复洗涤后从诱发的皱纹中恢复或保持光滑表面外观的能力。布料折皱回复能力,或者穿着、反复洗涤后保形能力是影响衣服外观的重要因素之一。工厂常用 AATCC 测试方法 128 "织物的折皱回复"作为织物折皱回复评估方法。

　　·起球评估

　　服装的外观质量还受到织物表面起毛起球的影响。毛球在织物表面形成分四个阶段:绒毛形成、纠缠、生长和磨掉。服装穿着过程中纺织面料上形成的毛球和其他表面变化(如起毛)会形成难看的外观。一些合成纤维在这方面特别严重,高强力的合成纤维将绒球挂在织物表面,不会像较弱的天然纤维那样使它们脱落。织物的抗起球性常在实验室的测试仪上进行翻滚、刷磨或揉搓测试以模拟穿着。样本与标准样本(实物或照片)进行比对,以确定样本对应于 5(无起球)–1 级(严重起球)之间起球等级。

·反复洗涤后的表面光洁度

AATCC 测试方法 12419 用于评估反复洗涤后平面织物的外观光洁度。测试程序和评估方法与上述方法几乎相同,只是试样制备和标准副本有所不同。

·缝合外观评估

缝合外观的视觉评估是在标准观察条件下接缝与标准样照进行对比。美国纺织化学和染色协会(AATCC)、美国材料测试协会(ASTM)、国际标准组织(ISO)和日本工业标准(JIS)都建立了各自的目视评估标准和程序。

·褶皱保持性评估

为了保持良好的服装外观,经反复洗涤的服装(尤其是裤子),其压痕应保留。AATCC 测试方法 88C27 用于评估织物褶皱保持能力。该方法是将褶皱织物样品进行标准洗涤,再在标准光源和观察区与参考标准对比样品的外观。标准提供了手洗或机洗、机洗过程和温度设置以及干燥程序设置的选择。沿织物长度、宽度方向切割三块织物样本,分别进行制样、压制、评级。AATCC 褶皱保持标准样本分为五个级别。

·成品服装外观保持性评估

由于织物尺寸稳定性和压制性能差、服装制造工艺差以及运输过程中的不利因素,服装外观会变差。羊毛服装的这个现象尤其突出。因此,国际羊毛秘书处日本分会提出了一种测试方法,用于评估男士西服在终端熨烫后和销售前的外观保持性。该测试方法将服装在一定的温湿度条件下放置一段时间,然后检查外观的变化。

Lesson 55   Textile recycling

A throw-away society with the realization that natural resources are threatened is a vivid illustration of the perplexing problem of contemporary lifestyle. As we consider the case of textile and apparel recycling, it becomes apparent that the process impacts many entities and contributes significantly. Because textiles are nearly 100% recyclable, nothing in the textile and apparel industry should be wasted. Harley Davidson jackets go to Japan, neckties to Vietnam, raincoats to London, cotton shirts to Uganda, sleepwear to Belize, shoes to Haiti, Levi's are coveted all over the world, and worn-out promotional T-shirts are made into shoddy or wiping rags. In 2021, it was projected there would be a 1% increase in world fiber consumption, which equals 2 million tons per year.

The textile and apparel recycling effort is concerned with recycling, recyclability, and source reduction of both pre-consumer and post-consumer waste. Although textiles seldom earn a category of their own in solid waste management data, the Fiber Economics Bureau reports that the per capita consumption of fiber in the USA is 80 pounds with over 40 pounds per capita being discarded per year. A report shows that China has surpassed the USA, making China the number one fiber consumer in the world. This report points out that China will continue to have the fastest growing fiber consumption market for the next ten years.

1.The problem of over-consumption

To compound the notion of over-consumption is the notion of fashion itself. The very definition of fashion fuels the momentum for change, which creates demand for ongoing replacement of products with something that is new and fresh. In addition, fashion has reached its tentacles beyond apparel to the home furnishings industry. Thus, fashionable goods contribute to consumption at a higher level than need. But without the notion of fashion, the textile, apparel, and home furnishings industries would realize even more vulnerability in an environment that is already extremely competitive. Apparel companies in the USA today have continual fashion "seasons" that constantly capture consumer interest as they stimulate sales and profits.

2.The textile recycling industry

The textile recycling industry is one of the oldest and most established recycling industries in the world; yet few people understand the industry, its myriad players, or reclaimed textile products in general. Throughout the world, used textile and apparel products are salvaged as reclaimed textiles and put to new uses. Textile recycling industry is able to process 93% of the waste without the production of any new hazardous waste or harmful by-products. The Council for Textile Recycling has indicated that virtually all after-use textile products can be reclaimed for a variety of markets that are already established. Even so, the textile recycling industry continues to search for new viable value-added products made from used textile fiber.

生词与词组

1.throw-away ['θrəuəˌwei]n. 废弃；废品
2.vivid illustration 生动的插图

3.perplexing [pəˈpleksɪŋ]*adj.* 使人困惑的

4.contemporary lifestyle 当代生活方式

5.promotional [prəˈməʊʃənl]*adj.* 促销

6.shoddy [ˈʃɒdi] *adj.* 劣质的；冒充的

7.wiping rag 抹布

8.pre-consumer 消费前

9.post-consumer 消费后

10.over-consumption 过度消费

11.ongoing [ˈɒngəʊɪŋ] *adj.* 不断发展的

12.tentacle [ˈtentəkl]*n.* 触角

13.vulnerability [ˌvʌlnərəˈbɪləti]*n.* 脆弱

14.competitive [kəmˈpetətɪv]*adj.* 竞争的

15.salvage [ˈsælvɪdʒ]*v.* 利用

16.hazardous waste 危险废物

17.harmful by-product 有害副产品

18.after-use textile 用后纺织品

19.value-added product 新增值产品

译文

第55课　纺织品回收

自然资源受到威胁,废弃社会是当代生活方式令人困惑的生动例证。当我们了解纺织品和服装回收现状时,清晰发现该过程影响了许多团体及其大量的产出。纺织品几乎100%可回收,因此纺织和服装行业不应浪费任何东西。哈雷戴维森的夹克去日本,领带去越南,雨衣去伦敦,棉衬衫去乌干达,睡衣去伯利兹,鞋子去海地,李维斯牛仔裤风靡全球,旧的促销T恤被制成次等布充好或擦拭布。2021年,预计世界纤维消费量将增加1%,相当于每年200万吨。

纺织品和服装回收成效与回收、可回收性以及消费前和消费后废物的资源减少相关。尽管纺织品在固体废物数据中少有自己的类别,但纤维经济局报告称美国的人均纤维消费量为80磅,每年人均丢弃超过40磅。一份报告显示,中国已经超过美国,成为世界上第一大消费纤维国。

该报告指出,未来十年中国仍将是增长最快的纤维消费市场。

1. 过度消费问题

时尚概念本就与过度消费概念相重合。时尚概念推动了变革,创造了不断用新鲜产品更换的需求。此外,时尚的触角已从服装延伸到家居用品行业。因此,时尚品对消费的贡献远高于需求。但无时尚的概念,在竞争激烈的环境中纺织、服装和家居用品行业将更加脆弱。现在美国的服装公司有连续的时尚"季",不断吸引消费者的兴趣以促进销售和利润。

2. 纺织品回收行业

纺织品回收行业是世界上最古老、最成熟的回收行业之一;然而很少有人了解该行业、众多行业参与者以及常见的回收纺织品。世界各地,用过的纺织品和服装产品均可作为回收纺织品进行回收使用。纺织品回收行业可处理 93% 的废物,而不会产生任何新的危险废物或有害副产品。纺织品回收委员会表示,几乎所有的用后纺织品都可被各种市场进行回收。即便如此,纺织品回收行业仍在继续寻找由纺织纤维制成的新增值产品。

References

[1] Hearle, J. W. S. *Handbook of textile fibres*[M]. 4th ed. Cambridge: Woodhead publishing limited, 1984.

[2] Saville, B. P. *Physical testing of textiles*[M]. 4th ed. Cambridge: Woodhead publishing limited, 1999.

[3] Annex, B. H. *Handbook of technical textiles*[M]. Cambridge: Woodhead publishing limited, 2000.

[4] Weber, K. P. *Knitting technology*[M]. 3rd ed. Cambridge: Woodhead publishing limited, 2000.

[5] Hu, J. L. *3-D fibrous assemblies*[M]. Cambridge: Woodhead publishing limited, 2008.

[6] Russell, S. J. *Handbook of nonwovens*[M]. Cambridge: Woodhead publishing limited, 2007.

[7] Simpson, W. S., Crawshaw, G. H. *Wool: Science and technology*[M]. Cambridge: Woodhead publishing limited, 2002.

[8] Gordon, S., Hsieh, Y. L. *Cotton: Science and technology*[M]. Cambridge: Woodhead publishing limited, 2007.

[9] Hearle, J. W. S., Hollic, L., Wilson, D. K. *Yarn texturing technology*[M]. Cambridge: Woodhead publishing limited, 2001.

[10] Clark, M. *Handbook of textile and industrial dyeing*[M]. Cambridge: Woodhead publishing limited, 2011.

[11] Horne, L. *New product development in textiles*[M]. Cambridge: Woodhead publishing limited, 2012.

[12] Lawrence, C. A. *Advances in yarn spinning technology*[M]. Cambridge: Woodhead publishing limited, 2007.

[13] Gandhi, K. L. *Woven textiles*[M]. Cambridge: Woodhead publishing limited, 2012.

[14] 王善元,于修业. 新型纺织纱线(英文版)[M]. 上海：东华大学出版社,2007.

[15] 李思龙,沈梅英. 纺织服装基础英语 [M]. 北京：中国纺织出版社,2017.

[16] 卓乃坚. 纺织英语 [M]. 3 版. 上海：东华大学出版社,2017.